CHEMICAL ENGINEERING METHODS AND TECHNOLOGY

VANADIUM

CHEMICAL PROPERTIES, USES AND ENVIRONMENTAL EFFECTS

CHEMICAL ENGINEERING METHODS AND TECHNOLOGY

Additional books in this series can be found on Nova's website under the Series tab.

Additional E-books in this series can be found on Nova's website under the E-book tab.

BIOCHEMISTRY RESEARCH TRENDS

Additional books in this series can be found on Nova's website under the Series tab.

Additional E-books in this series can be found on Nova's website under the E-book tab.

CHEMICAL ENGINEERING METHODS AND TECHNOLOGY

VANADIUM

CHEMICAL PROPERTIES, USES AND ENVIRONMENTAL EFFECTS

VENIAMIN N. BARANOVA
AND
ARKADY V. FORTUNATOV
EDITORS

Nova Science Publishers, Inc.
New York

For permission to use material from this book please contact us:
Telephone 631-231-7269; Fax 631-231-8175
Web Site: http://www.novapublishers.com

Library of Congress Cataloging-in-Publication Data

Vanadium : chemical properties, uses, and environmental effects / editors, Veniamin N. Baranova and Arkady V. Fortunatov.
p. cm.
Includes bibliographical references and index.
ISBN 978-1-62081-600-4 (hardcover)
1. Vanadium compounds. I. Baranova, Veniamin N. II. Fortunatov, Arkady V.
TP245.V2V36 2011
546'.522--dc23
2012013585

Published by Nova Science Publishers, Inc. †New York

CONTENTS

PREFACE

In this book, the authors present topical research from across the globe in the study of the chemical properties, uses and environmental effects of vanadium. Topics discussed include vanadium in biology and its contributions on the environmental cycles of chemical elements; ferritic stainless steel protected vanadium alloy as cladding material for fast reactors; vanadium oxides for solar control applications; global scientific production of vanadium research; and soft solution processing of nanoscaled lithium vanadium oxides as cathode materials for rechargeable lithium ion batteries.

Chapter 1 - Vanadium is essential for a considerable number of species from different kingdoms of eukaryotes and prokaryotes, but is less required for organisms belonging to higher ranks of evolutionary tree of life. Curiously, vanadium occurs in significant amounts in minus recent samples of coal and crude oil, but usually it is present at lower levels in the more modern ones, suggesting this metal has been more important for ancient life forms.

Vanadium has at least six different roles in living organisms: (1) in various seaweeds, lichen and terrestrial fungi, as haloperoxidases (2) in some diazotrophs, as alternative nitrogenase, (3) in few bacteria, as nitrate reductase molybdenum-free or as substrate for primary electron acceptor in dissimilation or respiration functions, (4) in ascidians, (5) in some marine worms, and (6) in few Amanita fungi, as amavadine. The cases (4)-(6) include the vanadium accumulators concerning their environments that are more significant for (4) and (5). For all these species it occurs at oxidation states within range three to five, being the upper limit the more widespread as result of substantial number of genera containing vanadium-haloperoxidases. The essentiality of vanadium for humans is not known, although for some mammals (such as, goats and rats) it is indispensable, but in these cases the biomolecules containing it have not yet been identified. Note that vanadium in biology is associated probably in all cases with redox reactions. At least one vanadium enzyme (nitrogenase) participates in modern nitrogen cycle. The relevance of this non-metal as restraining biomass formation gives to vanadium a considerable contribute on this process. Furthermore, its nitrogenase comprises a specific capacity of CO-reducing, suggesting that a precursor of this current enzyme were associated to ancient carbon cycles. Vanadium-haloperoxidases are linked to halogen cycles (except fluorine that is not substrate of these enzymes), including in production of volatile organic halogenated compounds, which are important in regulation of ozone levels in the atmosphere. Vanadium metallobiomolecules can also participate as buffer potential regulators in life forms, as protectors against predators by its accumulation in parts of the bodies or involving in synthesis of compounds difficult to be metabolized by other organisms. Vanadium has biological roles participating in the regulation of several ecosystems. However, accidental release of crude oil (containing high levels of vanadium, such as, from Venezuela) or its use as fuel without pre-treatment can disturb ecosystems, but this subject is out of this chapter.

Chapter 2 - The most important elements in microalloyed steels are niobium and vanadium. Nb is used in low carbon steels known as HSLA (High Strength and Low Alloy) steels. V is used in HSLA steels as well as in medium and high-carbon steels. Compared to Nb, the advantage of V is the very low solubility temperature of carbides, so the nanoprecipitates are usually vanadium nitrides. However, Nb cannot be used in medium and high-carbon microalloyed steels because the solubility temperature of niobium carbides and carbonitrides is very high and these precipitates do not dissolve at reheating temperatures prior to thermomechanical treatment. In this work, an analysis of V nanoprecipitates in thermomechanically-processed steels is carried out by means of scanning and transmission electron microscopy. It is shown that these particles have a mean size of a few nanometers (< 30 nm) due to their low solubility temperature. In consequence, and according to Orowan equation, VN precipitates cause a more remarkable increase in yield strength than NbCN precipitates. Another benefit of V microalloyed steels introduced in this work is the intragranular nucleation of ferrite and bainite in austenite grains existing before phase transformation. VN particles have low interfacial energy with respect to ferrite but relatively high interfacial energy with respect to austenite for the (001)VN boundary compared with other particles or inclusions. Such advantages promote the intragranular nucleation of ferrite and bainite. This work shows that V is an essential alloying element in high strength (> 1000 MPa) microalloyed steels with a ferrite-pearlite microstructure, as well as in bainitic steels. In low carbon steels, with ultimate strength near 600-700 MPa, Nb is also very interesting as a microalloying element. It is used in lower amounts, due to its higher solubility temperature and consequently its higher no-recrystallization temperature, which allows a more severe strengthening of austenite and a higher increase in dislocation density. Nevertheless, the intragranular nucleation of ferrite and bainite in V microalloyed steels make this element the best choice, provided that optimal additions are applied. Results about the static recrystallization of austenite during hot rolling and the recrystallization-precipitation-time-temperature (RPTT) diagrams will be also presented in this work. RPTT diagrams display interesting results about the minimum time for precipitation, the recrystallized volume fraction that favors precipitation and the optimal rolling conditions to obtain a very fine ferritic grain size. On the other hand, the increase in yield strength in high strength steels due to the presence of VN precipitates and the improvement in toughness in bainitic steels will be shown. An optimal V/N/Mn ratio will be established. The amount of V used in microalloyed steels is usually between 0.10 and 0.20 %, which means there is an important industrial market for this element.

Chapter 3 - The conditions of cladding structural materials work of promising fast reactors, as well as opportunities for the different classes materials use are analyzed. The paper demonstrates the advantages of vanadium base alloys with titanium, and chromium. Showing the principal foundations of structural materials technology for fuel rod claddings of fast reactors under conditions where corrosion resistance is provided by type Fe-17Cr ferritic stainless steel, which has the best corrosion resistance in atmospheric conditions and in pond, and the radiation resistance and heat resistance – by vanadium base alloys.

Chapter 4 - Solar control window coatings are those which can respond in a desired way to an external stimulus. These are materials such as the vanadium dioxide that can control the amount of daylight and solar heat that can pass through the windows of buildings with temperature.

Chemical vapour deposition is an attractive method for thin film deposition since it is capable of achieving high deposition rates. The properties of the as-deposited samples can be simply tuned by altering the vapour composition and flow in the coating zone. This can be beneficial for the thermochromic coatings since stoichiometric vanadium dioxide coatings are required. Finally, when CVD is performed at atmospheric pressure, it has the advantage of a relatively simple apparatus (no vacuum system required), which can be easily integrated into float-glass production lines.

Results obtained demonstrate the excellent thermochromic properties of undoped vanadium dioxide coatings as illustrated by switching temperature as low as 51 oC. Finally, from the in-situ monitoring analysis it can be possibly concluded that in the absence of oxygen, anhydride species may form with some evidence of ester species. While, in the presence of oxygen, there is indication for the enhanced production of CO and the suppression of the C-H species.

Chapter 5 - The chapter focuses on the progress made in the field of vanadium research over the last two decades. Many bibliometric studies have appeared in the scientific literature which focuses attention on the global trend for research productivity in the most varied subjects but at the moment there are no studies analyzing the world literature on vanadium. The purpose of this study is to identify research output, and research areas, as well as descriptively and quantitatively characterize the field of vanadium through publication trend analysis over the last twenty years. The growth of the literature on vanadium was found exponential in nature. During 1990-2011 a total of 35594 outputs were published by scientists in this field in various countries. The average numbers of research output per year were 1618. United State is the top producing country with 6224 outputs (17.5%) followed by The People's Republic of China with 4599 outputs (12.9%) and Japan with 3672 outputs (10.3%). The Chinese absolute numbers of research outputs continued to increase over time. The rise of The People's Republic of China has been especially notable, overtaking United States and Japan in terms of its research output in recent years. University of Tokyo topped the list with 375 research outputs (1.05%) followed by Russian Academy of Science with 349 outputs (0.98%) and CNRS Centre National de la Recherche Scientifique with 336 outputs (0.94%). About nine-tenths of the documents on vanadium were published in English language. The most preferred journals by the scientists in this topic were Inorganic Chemistry with 586 outputs (1.65%) followed by Physical Review B Condensed Matter and Materials Physics with 456 (1.28%) and Applied Catalysis A General with 444 outputs (1.25%). The most preferred subject area was Materials Science with 12925 outputs (36.31%). The high frequency keywords were Vanadium (14430 documents, 40.54%), Vanadium Compounds (6662 documents, 18.72%) and Vanadium Alloys (3741 documents, 10.51%). The top-five chemical elements in the keyword list were Chromium (2716 documents, 7.63%) followed by Titanium (2352 documents, 6.61%), Nickel (2331 documents, 6.55%), Aluminum (2184 documents, 6.13%) and Copper (2135 documents, 5.99%).

Chapter 6 - Vanadium (V) is a transition element, is a member of Group 5, and is the 23rd element of the Periodic table. Andres Manuel Del Rio was the first chemist to provide the idea of this new element in 1801. But it was discovered by Nils Sefstrom, a Swedish chemist in 1830. Pure V is a bright sliver-white, soft and ductile metal and 22nd most abundant element in the earth's crust. V has become subject of interest amongst nutritionists since the discovery that various marine species have this metal as an essential element. Although most food contains low amount of V (<1 ng/g), food is the major source of exposure to V for general population.

V may be beneficial and possibly essential in humans, but certainly essential for some living organisms. The usual oxidation states of the vanadium are V (III), V (IV) and V (V). The V ions play a role in biology as counter ions for various proteins, DNA and RNA.

V is widely distributed on Earth; however its role as a micronutrient in humans is not yet established. Humans are exposed to V mainly through the polluted atmosphere from combustion products of vanadium-bearing fuel oils, fumes and dust. Food contains a very low content of V usually below 1 ng/g. V enters the organism by inhalation, skin and gastrointestinal tract and accumulates mainly in the liver, kidney, spleen, bones and to a lesser extent in lungs and brain. The structural role is often manifested by the maintenance of various biological structures, whereas a functional role is to bring key reactivity to a reaction center for a protein. V ions have many structural roles reflected by its structural and electronic analogy to phosphorus. In addition, the V ion is an enzyme cofactor, and is found in certain tunicates and possibly mammals. During the last few decades, the facade of V as a 'slightly' toxic and carcinogenic element eventually ratified to an essential trace element with anti−diabetic and anti-carcinogenic properties. Neurotoxic effects of V are not well understood yet, but it is known that acute poisoning in animals by ingestion or inhalation of V compounds leads to nervous disturbances, paralysis of legs, respiratory failure, convulsions, bloody diarrhea, and death. V penetrates the blood–brain barrier and induces neurochemical alterations: noradrenaline, dopamine, and 5-hydroxytriptamine altered levels and an inhibitory effect on the uptake and release of noradrenaline were found in the rat brain during V poisoning. It is well established that vanadate (V^{5+}) and vanadyl (V^{4+}), the V oxidation states of biological interest, may cause a number of adverse toxic effects in mammals depending on circulating V levels. Industrial exposure of metal workers to vanadium pentoxide (V_2O_5) leads to cardiovascular diseases and a variety of symptoms involving central nervous system (CNS), gastrointestinal and respiratory systems. Moreover, it has been suggested that raised tissue levels of V may be of etiological importance in manic-depressive illness. Serum V levels in adults with depression were significantly greater than those of controls, regardless of the type of depression, tremor and impaired conditioned reflexes, as well as congestion of brain spinal cord. Moreover, reduced cognitive abilities in humans chronically exposed to this element were found.

Chapter 7 - Lithium vanadium oxides have been extensively studied because of their possible application as a cathode material for rechargeable lithium batteries. Due to their low cost, they are one of the promising substitutes for the expensive LiCoO2 cathode presently commercially used. Lithium vanadium oxides including γ-LiV2O5 and LiV3O8 have been prepared by soft solution methods in this study. In the first part of this work, γ-LiV2O5 nanorods have been prepared directly by a simple solvothermal method using ethanol as a solvent, which also serves as a reducing agent. The g-LiV2O5 nanorods with diameters of 30-40 nm obtained at 160 oC shows a larger capacity of 259 mAh/g in the range of 1.5 - 4.2 V, and its capacity remained 199 mAh/g after 20 cycles. In the second part, LiV3O8 nanorods have been obtained by a novel hydrothermal-based two-step method. The LiV3O8 sample treated at 300 oC shows a poor crystallinity while a specific capacity of 302 mAh/g in the range of 1.8 - 4.0 V, and its capacity remained 278 mAh/g after 30 cycles. It indicates that the lithium vanadium oxide nanorods prepared by the above methods have potentiality to be used as cathode material in rechargeable lithium ion batteries.

In: Vanadium
Editors: V. N. Baranova and A. V. Fortunatov

ISBN 978-1-62081-600-4

Chapter 1

VANADIUM IN BIOLOGY: ITS CHEMISTRY AND SOME OF ITS CONTRIBUTIONS ON ENVIRONMENTAL CYCLES OF CHEMICAL ELEMENTS

José A. L. da Silva*

Centro de Química Estrutural – Complexo I, Instituto Superior Técnico, Universidade Técnica de Lisboa, Lisboa – Portugal

ABSTRACT

Vanadium is essential for a considerable number of species from different kingdoms of eukaryotes and prokaryotes, but is less required for organisms belonging to higher ranks of evolutionary tree of life. Curiously, vanadium occurs in significant amounts in minus recent samples of coal and crude oil, but usually it is present at lower levels in the more modern ones, suggesting this metal has been more important for ancient life forms.

Vanadium has at least six different roles in living organisms: (1) in various seaweeds, lichen and terrestrial fungi, as haloperoxidases (2) in some diazotrophs, as alternative nitrogenase, (3) in few bacteria, as nitrate reductase molybdenum-free or as substrate for primary electron acceptor in dissimilation or respiration functions, (4) in ascidians, (5) in some marine worms, and (6) in few *Amanita* fungi, as amavadine. The cases (4)-(6) include the vanadium accumulators concerning their environments that are more significant for (4) and (5). For all these species it occurs at oxidation states within range three to five, being the upper limit the more widespread as result of substantial number of genera containing vanadium-haloperoxidases. The essentiality of vanadium for humans is not known, although for some mammals (such as, goats and rats) it is indispensable, but in these cases the biomolecules containing it have not yet been identified. Note that vanadium in biology is associated probably in all cases with redox reactions.

At least one vanadium enzyme (nitrogenase) participates in modern nitrogen cycle. The relevance of this non-metal as restraining biomass formation gives to vanadium a

* Centro de Química Estrutural – Complexo I, Instituto Superior Técnico, Universidade Técnica de Lisboa, Av. Rovisco Pais, 1, 1049-001 Lisboa – Portugal, e-mail: pcd1950@ist.utl.pt

considerable contribute on this process. Furthermore, its nitrogenase comprises a specific capacity of CO-reducing, suggesting that a precursor of this current enzyme were associated to ancient carbon cycles.

Vanadium-haloperoxidases are linked to halogen cycles (except fluorine that is not substrate of these enzymes), including in production of volatile organic halogenated compounds, which are important in regulation of ozone levels in the atmosphere.

Vanadium metallobiomolecules can also participate as buffer potential regulators in life forms, as protectors against predators by its accumulation in parts of the bodies or involving in synthesis of compounds difficult to be metabolized by other organisms.

Vanadium has biological roles participating in the regulation of several ecosystems. However, accidental release of crude oil (containing high levels of vanadium, such as, from Venezuela) or its use as fuel without pre-treatment can disturb ecosystems, but this subject is out of this chapter.

INTRODUCTION

Vanadium is a first-row transition metal belonging to group 5 of the periodic table and has the atomic number 23. In Nature, this metal is composed predominantly by stable isotope ^{51}V, which is present with a relative natural abundance of 99.75%, and by another radioactive isotope ^{50}V that has a very long half-live. However, several other radioisotopes are known, which have typically short half-lives, being the majority of these obtained in particle accelerators or in nuclear reactors. In Earth, this metal has been identified in more than sixty different minerals, in which vanadium is usually bound to oxygen and in structures containing other metals [1,2].

Vanadium chemistry is diverse and the formal oxidation state of its compounds varies from -3 to 5, being the most common the higher three adjacent oxidation states (from 3 to 5). In aqueous solutions, this metal can occur at oxidation states from 2 to 5 and each one have a different color, as consequence of their different electronic configuration in the ground state. Additionally, it can produce for its higher oxidation states (4 and 5) oxygenated species that can form polymeric structures, which have different compositions dependent of vanadium concentration in solution, its oxidation state and pH values [2]. Furthermore, it is present in compounds with several non-metal donor atoms coordinated to metal, including in several organometallic complexes. Vanadium can be a constituent of polynuclear metal centers (clusters) usually with donor atoms doing interactions through a bridging ligand. Its clusters may have as metal only vanadium or can be together with other metals, and for more than one vanadium atom in the structure, they can have different oxidation states [2]. The coordination chemistry of this metal has variable coordination number and additionally, vanadium, in general, can behave as Lewis acid, due to its valence orbitals are electron-poor. These characteristics and also its diversity of oxidation states and donor atoms bounded to it, permit this metal to act as catalyst for several chemical reactions [2].

In spite of its varied chemistry, the main technological appliance of vanadium is in alloys above all, with iron, allowing its presence in the steels lightness and simultaneously increasing tensile strength. Vanadium has been used, as catalyst, for instance, in manufacturing of sulfuric acid, in reduction of industrial nitrogen oxide and sulphur emissions, being its compounds also applied as pigments in production of ceramics, as glass coating, in the nuclear power industry and in the health sciences [1,2].

GENERAL FEATURES OF VANADIUM IN NATURAL ENVIRONMENTS

Vanadium occurs in different geospheres (*i.e.*, lithosphere, hydrosphere, atmosphere and biosphere, which includes the global sum of all ecosystems) of the Earth, having the metal dynamic transfer flows among them, which comprise also, in biosphere, the involvement of some living organisms.

The chemical properties of vanadium control the incidence of these processes, although they are not independent of the presence of other chemical elements. Moreover, the natural erosion and the leaching may also be relevant routes for its natural transfer in biogeochemical cycles.

In reservoirs (compartments of our planet in which a material can be stored) of the Earth where vanadium has more significant levels are the crust and the sea-water, although is by atmosphere that this metal has its fastest spreading over the world, occurring in this geosphere compartment as small airborne particles that are mostly originated by human activities.

This metal is the 19th most abundant chemical element in the Earth's crust with a considerably larger amount than other more used metals in current human applications (zinc, nickel or copper)[1], while, in sea-water it is the 2nd most transition metal in abundance, after molybdenum, but is much higher than iron. However, its average concentration in the oceans is far less than in the crust. Vanadium is considered a quasi-conservative in its marine distribution, although it is minus in coastal waters, having a residence time in oceanic waters of 0.05 My [1,3-5].

The predominant oxidation state of vanadium in both aquatic and geological environments is five, typically as vanadates or vanadium pentoxide [5]. However, it is noted that in the oceans vanadium levels and its oxidation states may be different in benthic and pelagic zones, being usually in latter location vanadium more abundant and occurring in higher oxidations states than in the former ones [6].

On the other hand, vanadium is a redox-sensitive trace-metal, forming diverse types of compounds in natural sediments as consequence, for instance, by alteration of dioxygen levels and by the presence of hydrogen sulfide in some milieus.

The variety of environmental conditions cause lower or higher incidence of this metal in distinct types of compounds, whose composition are dependent of the solubility of vanadium more or less oxidized species that are related with conditions where the sediments have been formed.

These characteristics of vanadium permit to identify the increase of oxygen molecular levels in some older Earth's geological periods or in transitions between anaerobic to aerobic environments (or vice versa), which were factors that affect relationships between this metal with biological systems and, as consequence, might have permitted evolutionary trends.

Taking into consideration vanadium capacity as redox-sensitive for various alterations in some milieus, its sediments can be employed as a proxy of oxic, hypoxic, anoxic or anoxic-sulfidic events, including for research of ancient redox chemistry and, as a particular case, in biological chemistry [4].

---

[1] Regardless of the relative abundance of vanadium in the Earth's crust, it is considered minus common metal than others more used in human activities, although the metals belonging to latter group are scarcer as natural resources comparing with vanadium. The explanation for this assertion is due to small scale production of this metal that is a consequence of its irregular distribution in Earth and also the typically low concentration of vanadium in its minerals [1].

General Features of Biological Chemistry of Vanadium

The requirement of a specific chemical element by a living organism for a biological role is dependent of its particular biochemical advantageous. In some cases the storage of a chemical element is needed previously of its metabolic employment. These circumstances occur chiefly in the cases of low abundance (particularity for living organisms without ability to access of materials of different origins, as typically occurs with life forms based on absence of mobility) or low availability of the chemical element. Subsequently, for its active incorporation in organism the transport from storage location is necessary by using a specific carrier that after its release from it, permit its utilization for a biological function. In general, the use of a chemical element is necessarily dependent of low-energy spending procedures (from uptake to its active role) and ensuring that its employ minimizes toxic effects in metabolism [7].

The number of chemical elements known as essentials for at least one living organism is roughly one-third of all occurring in periodic table that have at least one isotope non-radioactive. However, only a smaller group of chemical elements is essential to all organisms (probably less than half of all identify essential chemical elements), being vanadium crucial for a limited number of biological systems and it is included in group of oligoelements that contains mainly transition metals, despite some of these chemical elements are essential for great number (or all) living organisms, as is the case of iron [8].

Note that each chemical element can occur in more or less complex molecules or ions in organisms. Taking into account the specificities of each chemical element every organism has mechanisms for its internal level regulation (and its activity) into a narrow range and also for its uptake from the external environment, including the use of resistance factors for some chemical elements, to avoid disturbance on metabolism. It should be noted that the oxidation state of a chemical element may also be important in terms of its toxicity [8].

Vanadium typically occurs in low levels in living organisms and is essential for some species from different kingdoms of eukaryotes and prokaryotes, although it is less necessary for metabolism of the organisms belonging to the higher ranks of evolutionary tree of life. This metal is spanned from cyanobacteria (roughly originated at 3 billion years ago) to green seaweeds (*ca.* 0.8 billion years old) [3]. However, bearing in mind its essentiality for some mammals the upper limit is closer of our days [9]. Curiously, vanadium occurs in significant amounts in minus recent samples of coal and crude oil, but usually it is present at lower levels in the more modern ones, suggesting that this metal has been more intensively used by ancient life forms [10-12], probably as consequence of increase of availability of other chemical elements resulting from the formation of a more oxygenated atmosphere during ulterior geological history of Earth, which have allowed vanadium replacing in some biological roles (molybdenum is a plausible case of this event, which is in diagonal of vanadium in periodic table belonging to group 6 and participate, at least, in one metallobiomolecule that can alternatively contain vanadium – see below, Vanadium roles in living organisms) [8].

Vanadium essentiality in the early stages of evolution may be a consequence of its general ubiquity, its higher availability concerning several metal transition in that period and some of its chemical properties, such as, amphoteric behavior and capacity of occurrence in some oxidations states under physiological conditions [3,8].

The biological chemistry of vanadium is less varied than its laboratorial chemistry regarding diverse aspects, such as its range of oxidation states (from 3 to 5) or donor atoms to metallic center (less diverse in vanadium biomolecules), but it has some specificities, for example, in relation to unusual coordination features (see below, amavadine) or concerning some particularities of biological catalysts, giving them, for instance, more efficiency in comparison with their models [8]. Usually, internal distribution of vanadium in an organism is not homogeneous both concerning levels or oxidation states.

Note that the vanadium in biology have electrons in valence orbitals for all oxidation states except to higher, the five, in Earth its predominant isotope has odd nuclear number (^{51}V), and occurs in different oxidation states. These properties permit the employment of diverse techniques for characterization its biological compounds, although the isolation and purification without denaturation of its metallobiomolecules and, in some cases, its low amounts in organisms are a number of difficulties for an accurate structural and functional biological characterizations of vanadium in these systems.

The group of a chemical element in periodic table influences its available structures in natural environments. In case of vanadium, due to be a metal transition of left side of periodic table, its natural bioavailability is typically as anionic species, but in less oxidant environments it is also possible its occurrence as cationic species (in lower oxidation states). In view of vanadium uptake (or assimilation) is as ion, it is necessary for its entrance in the cells cross them at specific sites employing mainly pumps or using carriers [8,13].

In some cases vanadium in organisms is present in lower oxidation states than in external media and its reduction is needed by metabolic mechanisms to be possible this metal performs its biological role. Typically, vanadium is available as vanadate and chemical reduction of this chemical species has a significant reduction potential value, although it is dependent of pH value and terminal species of this metal in the chemical reaction. Each redox process is coupled with other, involving the chemical reduction route the addition of an electron or hydrogen atom to an atom or molecule, and concomitantly leading to oxidation of the other. Vanadate reduction is possible by oxidation of several substrates, which can lead to pernicious side reactions, such as, radicalar or toxic species for the organisms, but can allow the employ of these products, for instance, in defensive strategies against predators. Additionally, vanadate has some chemical similarities with phosphate, being the latter a main raw material of several crucial biomolecules present in all living organisms. These resemblances between both chemical species cause some inhibitory metabolic behaviors by vanadate in metabolism associated to phosphate [8]. Furthermore, the similarities of vanadate and phosphate related to charge at equivalent pH values can allow the uptake of the metallic anion by systems normally used for the non-metallic anion [8].

To be avoid the surpassing of the upper limit levels of vanadium accepted in the normal metabolism the organisms can develop different detoxification procedures to keep away from dangerous effects resulting from this metal hazardousness, such as, by changing oxidation state of metal, its complexation, or storage in specific sites of their biological structures. It should be noted, however, that the human achievements have become this metal more available, affecting the interactions between vanadium with living organisms and as consequence can lead to harmful effects. The technological applications of vanadium are associated to different activities, such as, mining, smelting and purification, which favor a significant diffusion of its compounds. Moreover, employing crude oil or coal containing high levels of vanadium, for example, as raw material in chemical industry or as fuel, allows the spreading of considerable amounts of vanadium residues as dust, with the metal typically as vanadium pentoxide [8].

Anthropogenic activities, even using vanadium as minor component, can lead to its toxicity in living organisms (due to metabolic competition with other chemical elements) and/or inducing chemical modifications in different environments (man's industrial activities have already exceeded the emissions of this metal from natural sources - resulting mainly from continental dust, marine aerosol, and volcanic emissions - to atmosphere [14]). Vanadium release in natural waters causes changes in its pH values or alterations in the availability of other chemical elements, provokes environment erosion, modifies soil composition, and originates ground and surface water contamination, which can also affect the biological systems. However, this chapter is about the role of living organisms in transfer of vanadium to in or out biosphere associated to the metabolism of living organisms and is also concerning some influences that this metal in its biological roles can bring on cycles of other chemical elements in the environments. In spite of anthropogenic contribution is a substantial influence in release and transfers of vanadium in and among all reservoirs of Earth this subject is beyond the purpose of the present chapter.

VANADIUM ROLES IN LIVING ORGANISMS

Although vanadium is an essential oligoelement for some living systems, it is also known a smaller group of organisms with the capacity for an unusual vanadium accumulation in their bodies concerning their around environments, having this feature significant relevance for some organisms living in the seas (ascidians and few worms, see below). Usually, vanadium occurs in marine organisms in range of ng/g, but this value is considerable higher for the two examples presented [15,16].

In general, the accumulation of typically trace-level chemical elements by organisms is associated to protection against disadvantageous environmental conditions. This capacity can be used as an improvement for passive protection against predators, prevent infections caused by microorganisms, increase antifouling defenses, or as consequence of a necessity resulting from its metabolism or of a symbiotic association[2], which can permit, such as, the accumulation of a chemical element for the partner, as an access of unusual resources or reducing the contact with hazardous substances. The accumulation of a specific chemical element in some biological structures can also be a path to avoid biological interferences of other available chemical elements [17,18].

After these general comments about the organisms with ability for accumulating chemical elements, it is resumed the subject related to vanadium and its known biological roles. Vanadium has at least six different roles in living organisms: (1) in some seaweeds, lichen and terrestrial fungi, as haloperoxidase (2) in diazotrophs (prokaryotes of some genera), in biological fixation of nitrogen (3) in bacteria, at least as primary electron acceptor, and also in some accumulators of this metal, (4) in ascidians, (5) in some marine worms, and (6) in few *Amanita* fungi. On the other hand, this is essential for some mammals, however,

[2] This type of relationship is known across taxonomic ranks and can be crucial for the contributors; habitually, it involves mutual recognition and an important coordination of the partners at morphological and physiological levels, needing a cellular and molecular communication among the components of symbiosis.

the biological molecules containing vanadium are not identified. In the set of all species this metal in biological systems occurs in various oxidation states within range three to five. Note that this metal in biology is associated probably in all cases with redox reactions.

In the following paragraphs additional aspects of each case will be mentioned.

(1) In some seaweeds (from several genera), lichen and terrestrial fungi (isolated from plants and soils, belonging to the dematiaceous hyphomycetes and from different genera) [19] and bacteria [20] the active center of haloperoxidases contains vanadium, which catalyze the reactions presented in equations 1 and 2. The same type of reactions are also catalyzed by a haloperoxidase with iron in active center that is included in a hemic system, being this variety of enzyme dispersed for different types of organisms [21]. However, hemic haloperoxidases differ strongly from the vanadium-dependent enzymes in substrate specificity and also in molecular properties [21,22]. Both types of haloperoxidases are peroxidases (*i.e.* using peroxide as substrate) that mediate the oxidation of halides (*i.e.* anions of the chemical elements of group 17 of periodic table and typically their chemical occurrence in Earth biological environments - only chloride, bromide and iodide are substrates, but fluoride is not due to its high oxidation potential can not be oxidized in these conditions) by hydrogen peroxide according to the following reaction:

$$H_2O_2 + X^- + H^+ \rightarrow H_2O + HOX \qquad \text{(eq. 1)}$$

$X^- = Cl^-$, Br^- or I^-

and in presence of the general nucleophilic acceptor, AH, occurs:

$$HOX + AH^+ \rightarrow H_2O + AX \qquad \text{(eq. 2)}$$

This reaction permits the formation of several halogenated compounds, including also the organic ones.

The classification of haloperoxidases, chloro-, bromo- or iodoperoxidase is attributed by most electronegative halide whose oxidation the enzyme is able to catalyze [19]. Note that the type of vanadium-haloperoxidases can be specific for different environments, while chloroperoxidases occurs in fungi from terrestrial milieus and bromo- and iodoperoxidases are present in marine organisms [19].

In vanadium-haloperoxidases the metal is present as vanadate in prosthetic group that is only coordinated to proteic component of enzyme by a nitrogen atom of a histidine residue, see Figure 1. The molecular weight of enzyme per vanadium atom is lower than 68,000 Da [19].

The active center of bromoperoxidases and chloroperoxidases are very similar and only a residue around it and not coordinated with vanadium is different (phenylalanine in chloroperoxidase of *Curvularia inaequalis* (Shear) Boedijn, 1907 [23] instead a histidine in bromoperoxidase of *Ascophyllum nodosum* (L.) Le Jolis [24] and iodoperoxidase differ from bromoperoxidase by alanine residue instead one of serine) [25] that may facilitating the selectivity concerning halide substrate as had been suggested by Almeida *et al.* [26]. Metal performance in the active site of enzyme binds hydrogen peroxide in a side-on fashion and acts as a Lewis acid allowing nucleophilic attack of an incoming halide that is oxidized into HOX [19], see Figure 1.

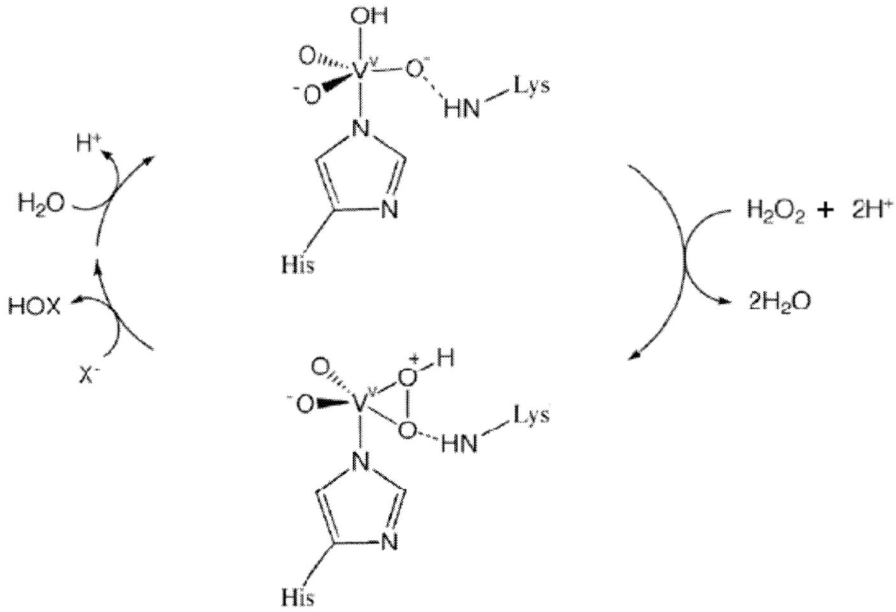

Figure 1. Schematic representation of vanadium-haloperoxidases mechanism. Adapted from [27]; His = histidine; Lys = lysine; X- = chloride, bromide or iodide.

These enzymes are very stable with the temperatures (more for the chloroperoxidases) and also for some solvent mixtures containing water and a non-aqueous solvent [19,22]. The former aspect can be important for seaweeds during the low tides and for fungi living in deserts (as is the case of some species above-mentioned containing the enzyme) due to reduce denaturation of haloperoxidases under intense sunshine.

Vanadium-haloperoxidases can present some reactive specific features for some substrates. So, chloroperoxidase of bacteria belonging to *Streptomyces* genus are highly stereoselective in biosynthetic route involving the formation of chlorinated polyketide-terpenoid antibiotics dubbed napyradiomycins [20], see Fig. 2.

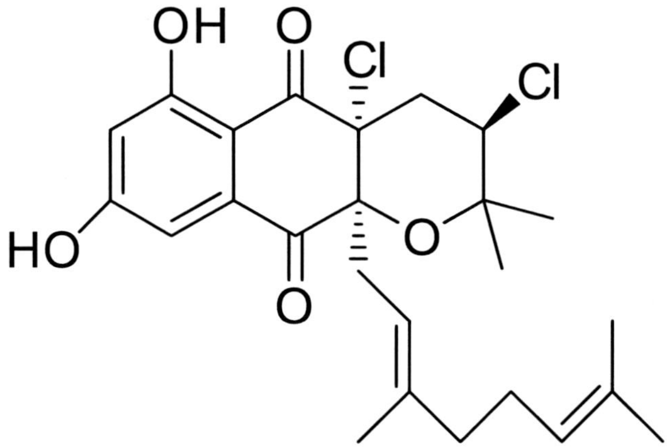

Figure 2. Example of a napyradiomycin.

On the other hand, in absence of halides vanadium-bromoperoxidases are also able to mediate other type of reactions, the enantioselectivity sulfoxidation of organic sulfides using hydrogen peroxide as oxidant, while chloroperoxidases produce racemic mixture of sulfoxides [28], see Figure 3.

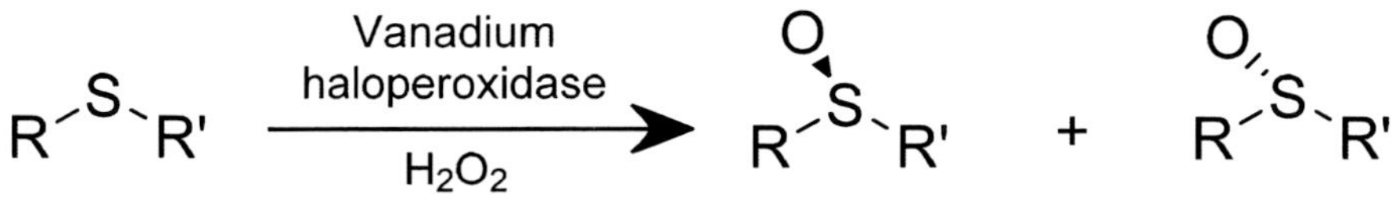

Figure 3. Sulfoxidation of organic sulfides catalyzed by vanadium-haloperoxidases; in some cases it is possible enantiomeric enrichment (see text). R = aromatic; R' = methyl.

Curiously, it was identified some structural and electronic similarities of vanadate with phosphate leading the former inhibit phosphatases and the latter vanadium-haloperoxidases. These features suggest that these enzymes can be evolved from the former ones [19].

The oxidation state five to vanadium is the most widespread in biology as result of substantial number of genera containing vanadium-haloperoxidases. Note that in natural environments vanadium occurs mainly in same oxidation state and usually as vanadate, which favoring low-energy spending for the uptake of the metal by organisms containing haloperoxidases dependent of vanadium.

The occurrence of haloperoxidases containing vanadium gives to this metal a considerable relevance in natural halogen cycles [see below, Vanadium in halogen (chlorine, bromine and iodine) cycles].

(2) The diazotrophs are microorganisms widespread among several genera of both prokaryotes domains (bacteria and archaea, but mainly in former one) and are able of carry out the biological fixation of nitrogen using nitrogenases, process crucial for all domains of life. Some diazotrophs are not directly associated with other organisms, but other develops mutualistic relationships with eukaryotes, especially with plants, to perform nitrogen fixation [29].

Nitrogenases have the capacity to fix nitrogen molecular (N_2) from atmosphere that is the largest pool of this chemical element in Earth. This process allows the production of ammonia (NH_3), a fundamental raw material for all biologic nitrogen, by successive reduction reactions. The most efficient overall reaction for this enzyme (molybdenum-nitrogenase, see below) needs large amounts of ATP:

$$N_2 + 8H^+ + 8e^- + 16MgATP \rightarrow 2NH_3 + H_2 + 16MgADP + 16Pi \qquad \text{(eq. 3)}$$

Nitrogen molecular is quite inert due to the strength of the multiple bonds between the two nitrogen atoms of the molecule requiring a catalyst (the enzyme nitrogenase is the unique with this capacity in biological systems) to permit its chemical conversion. However, there are some differences among all nitrogenases known, which can be dependent of ferredoxin or flavodoxin and have a cofactor containing molybdenum that may be replaced by vanadium or iron in case of scarcity of the former metal [29].

Vanadium alternative enzyme is found, for example, in *Azotobacter vinelandii* Lipman, *A. chroococcum*, *Anabaena variabilis* and *Rhodopseudomonas palustris*. It is expressed by specific genes that are dependent from environmental conditions (such as, levels of metals required for the enzyme active center and of the environmental temperature). The vanadium-nitrogenases have a molecular weight of roughly 250,000 Da, value different from molybdenum enzymes, including clusters that are redox centers with iron and the catalytic center contains probably, at least, one atom of vanadium in a cluster with Fe, S and homocitrate by subunity. Typically, vanadium-nitrogenases evolves more H_2, can release reduced intermediates of substrate and consumes more ATP in reduction of N_2 than molybdenum enzymes [29-32]. The structure of molybdenum component of its nitrogenase is determined, but not yet the corresponding vanadium-dependent. However, it was suggested [33], based on similarities with molybdenum enzyme, vanadium is in its enzyme at oxidation state three, but it is only a possibility due to the complexity of the cluster containing this metal [33].

In case of *A. vinelandii* require vanadium for nitrogenase biosynthesis, it can release pro-ligands, such as, tris-catechol protochelin or bis-catechol azotochelin, which are strong complexing agents for vanadate, the most abundant oxidation state of vanadium in natural environments, to uptake of the metal [34]. Additionally, the redox centers present in all nitrogenases have the capacity to reduce vanadium (V) at neutral value of pH, which can contribute for its conversion to active form in the enzyme [35].

The occurrence of a type of nitrogenase containing vanadium gives to this metal a significant relevance in natural nitrogen cycle (see below, Vanadium in nitrogen cycle)

(3) In some bacteria, vanadium may perhaps be present in active center of nitrate reductases molybdenum-free and certainly the mentioned metal, as vanadate, can occur as primary electron acceptor in metabolism of other species belonging of same domain.

The nitrate reductases are enzymes that catalyze the reduction of nitrate (NO_3^-) to nitrite (NO_2^-) by following reaction:

$$NO_3^- + 2e^- + 2H^+ \rightarrow NO_2^- + H_2O \qquad \text{(eq. 4)}$$

It is also known the capacity of these enzymes for nitric oxide (NO) biosynthesis, although there other metabolic reactions for its production, being this chemical species commonly accepted as signalling compound in several organisms [36].

In general, nitrate reductases can be assimilatory or dissimilatory, occurring also in eukaryotes, be present in diverse cellular locations and are typically molybdoenzymes [37].

Nitrate reductases lacking molybdenum, and containing vanadium, had been reported from the halo-alkaliphilic sulfur bacterium *Thioalkalivibrio nitratireducens* [38], but ulterior results show absence of vanadium [39] and also from the facultative anaerobic, chemolithotrophic bacterium *Pseudomonas isachenkovii*. This species has two different nitrate reductases, a soluble periplasmic and a membrane-bound, molybdenum-free dissimilatory not yet full characterized [40-42]. Probably, one of the nitrate reductases carry out vanadate reduction after nitrate reaction to nitrogen molecular has been almost concluded [3,40]. In case of bacterium anaerobic growth vanadate behaves also as primary electron acceptor and the organism secretes a protein (with molecular weight of 14 kDa) into its external environment with capacity of coordination up to 20 vanadyl ions (metal in oxidation state four) [40]. These proteins may be vanabins (see below, ascidians) as consequence of *P. isachenkovii* have been collected from intestinal contents of *Ascidia sydneiensis* var. *samea* Oka, 1935, that contains the mentioned protein [15]. The activity of this bacteria concerning vanadate can be a detoxification strategy for excessive levels of this anion in its habitat, which can disturb the activity of the biomolecules having or using phosphate derivatives due to some chemical similarities between both anions (see above, General features of biological chemistry of vanadium).

Taking into account *T. nitratireducens* recent structural results of its nitrate reductase [39] do not confirmed the presence of vanadium in enzyme, additional care is necessary for interpretation of the early results about the corresponding biomolecule of the *P. isachenkovii* [40-42], having been noted that this nitrate reductase comprises high sensitivity to inhibitors of hemic enzymes [43]. However, its capacity for reduction of vanadate may indicate rather a strategy of reducing the toxicity of this metal leading, for instance, to its mineralization, and can also employing vanadium in its dissimilatory functions [3].

The capacity of vanadium as primary electron acceptor in bacterial activity is a consequence of its different oxidation states allowed by its varied redox chemistry. Typically, vanadium occurs in natural environments its higher oxidation state five, as vanadate, giving its reduction chemical species lower oxidation states. Bacteria from different environments (including from deep-sea hydrothermal vents) belonging to genera such as *Shewanella* [44,45], *Pseudomonas* [46], *Vibrio* [47] and *Geobacter* [48] can use vanadates, by different pathways, as a primary electron acceptor in dissimilation or respiration. These activities give contributes by changing bioavailability of this metal in natural environments, permitting, such as, the decrease of its potential toxic effects for some organisms. Curiously, some strains of these bacteria can be associated with vanadium accumulators species [40,49,50], suggesting a defensive strategy against the toxicity of this metal.

However, the bacteria above-mentioned do not use exclusively vanadium as terminal electron acceptor, employing other chemical elements, including oxygenated anionic metalloids [44-49].

The next examples of vanadium in biology are related to species accumulators of this metal, starting by marine organisms.

(4) Ascidians, tunicates or sea squirts belong to the Ascidiacea class and to the phylum Chordata, corresponding to more than 2,800 marine invertebrate filter feeders species that inhabit in all oceans, in a wide variety of habitats, usually in shallow waters with salinities from 2.5% to 4.4%. Typically, they are rounded or cylindrical animals ranging from about 0.5 to 10 centimetres in size [51] and are able to concentrate vanadium up to 10,000,000 times higher than in seawater, as occurs with the species *Ascidia gemmata*, Sluiter, 1895 [52].

Vanadium is accumulated by a large number of ascidians predominantly at oxidation state three, but few amounts of the metal at oxidation state four are also present in their bodies. The metal reduction occurs during its assimilation and after it is stored in specific vacuolated blood so-called vanadocytes [53]. However, in these cells vanadium is not associated to dioxygen transport as previously it had been suggested [54].

Usually, vanadium is uptake at oxidation state five and its reduction to oxidation state four occurs in the cytoplasm. The subsequent reduction happens in vacuoles [55]. In metal accumulation process a specific vanadium-binding metallobiomolecules isolated from ascidians is involved, called vanabins. These proteins can coordinated several vanadium ions at oxidation state four, have a molecular weight between 12.5 to 16 kDa and are present almost exclusively in blood cells of ascidians. Each one has significant differences in amino acid residues composition suggesting a particular role for every vanabin. However, other proteins can be implicated in the vanadium accumulation process [15,56]. Bearing in mind that the main occurrence of vanadium in seawater is at oxidation state five, some reducing agents in ascidian blood cells have been proposed. Note that some species of ascidians live at environments with dioxygen depleting [57], where is more plausible the occurrence of vanadium at lower oxidations than five.

Despite the discovery of high levels of vanadium in ascidians has occurred more than one hundred years, having been the first case identify of this metal in biology, the current knowledge about biological role of vanadium in these organisms is still a rather intriguing subject [15,55].

(5) Some polychaetes (a group of marine worms), with special emphasis on tube-dwelling fan worms of the Sabellidae family (also known feather duster worms) and of the phylum Polychaeta, living typically in the intertidal zones around the world, can be vanadium accumulators at levels several orders of magnitude higher than those commonly found in the majority of aquatic organisms. Two examples are *Pseudopotamilla occelata*, Moore, 1919 [58], and the antarctic *Perkinsiana littoralis*, Hartman, 1967. Despite its similar capacity known for some ascidians (see above) to accumulate vanadium their phylogenetic relations with the phylum Chordata are distant [59].

This type of marine worms is sedentary, has approximately 15 cm in length and lives inside of a self build tube, which is associated to their passive defense. The tubes result from assemblage of detritus, such as, remains of animal tissues, sand, debris, and fragments of shells. Their heads can be present out of tubes and are chiefly hidden by appendages belonging to a branchial crowns that are usually expanding to outside for filter-feeding and respiration [16,60]. Vanadium concentrations are higher than 5,000 and 10,000 μg/g in branchial crowns. These structures contain great majority of vanadium in organism, despite their low contribution for overall weight of these species. The metal is stored at oxidation state three in vacuoles of the epithelial cells. The elevated levels of vanadium in branchial filaments of sabellids can act as chemical antipredatory defense in more exposed tissues [16]. Taking into account the dissolved vanadium species are predominantly at oxidation state five in upper levels of the oceans this type of polychaetes needs a system to reduce the metal, although these animals can live in less oxidant environments that allow the metal uptake at lower oxidation state.

(6) In few *Amanita* fungi vanadium content can exceed 400 times the value normally detected in other species of the same genus that has about 600 different species [13,61,62]. This feature is identified in native species of temperate and boreal regions of northern hemisphere, as well as high altitudes of lower latitudes, *Amanita muscaria* (L.:Fr.) Hook., and also in the its related species *Amanita regalis* (Fr.:Fr.) Mich. [61] (or *A. muscaria* var. *umbrina* Fr. [63]), as well as in *Amanita velatipes* Atkinson [61,62] or *Amanita gemmata* f. *amici* (Gill.) Gill., very close to *A. pantherina* (DC.:Fr.) Sécr., or to *A. muscaria* (L.:Fr.) Hook. [63]. Note that they are known different varieties of *A. muscaria* and can be also found currently in many regions of the southern hemisphere where it was introduced accidentally [64]. Despite these *A. muscaria* and related ones fungi that accumulate vanadium are terrestrial, they belong to a different phylum of the organisms containing haloperoxidases vanadium-dependent (see above), being yet the morphological features of the two groups very different.

A. muscaria general characteristics include typically a yellow, orange or red cap, which contains white spots, resulting from rests of provisional membranous tissue that wholly encloses immature fruiting bodies in its development. It occurs associated with various deciduous and coniferous trees forming a close, mutually beneficial relationship with the roots of its hosts [64]. Commonly, this fungus is named as fly-agaric by reason of its dormant effect in insects. It has yet psychoactive properties and by these features it has been used in different ethnic groups and integrate rituals in some popular religions, including for the believers to get into trance states, despite it contains several dangerous toxins. Historical documents with its representation have been found in different periods and from several locations, being in our days used, for instance, in illustrations of tales for infancy, in strip cartons, in Christmas decorations, comprising postcards [65].

Vanadium in *A. muscaria* was isolated as amavadine that is a particular complex with 401 Da. The metal presents in this biological compound a rare coordination, concerning vanadium and also first-row transition metals of periodic table, due to it has eight donor atoms. Additionally, the metallic center is a bare vanadium(IV) (also unusual for this chemical element that in this oxidation state occurs mainly as vanadyl, VO^{2+}), complexed by two pro-ligands of (S,S)-*N*-hydroxyimino-(2,2')-dipropionic acid [66], see Figure 4. The oxidized form of the complex (with the metal at oxidation state five) comprises minor structural differences comparing the complex with vanadium(IV) [66-68] and the value of the oxidation potential of the pair can support a function involving electron-transfer. It acts as an electron-transfer mediator in the oxidation of some biological thiols, [13,16,63,66-71] by Michaelis-Menten type mechanism, [69] or as a reducer of H_2O_2 and oxidant of water, [63], see Figure 5. Moreover, amavadine is an anion (for both oxidation states of vanadium) and has affinity by ammonium ion [72], which permits a strong interaction with amino acid residues with cationic side chains or with the N-terminus of proteins. These features lead to this biological compound as a plausible cofactor of oxidoreductases or to electron-transfer in metabolism, which may allow a defensive strategy in self-regeneration of damaged tissues or associated with its partnership in symbiosis [13,63]. However, at present there is no conclusive understanding on the function of amavadine in these fungi.

Additionally, concerning vanadium uptake is not known the mechanism by *A. muscaria*, but in preliminary results [73] it was observed the capacity of the pro-ligands of the biological complex to reduce vanadate. This possibility is also sustained in the amavadine has a similar coordination structure for both oxidation states, which leads to a plausible low-energy spending in all process of metal uptake.

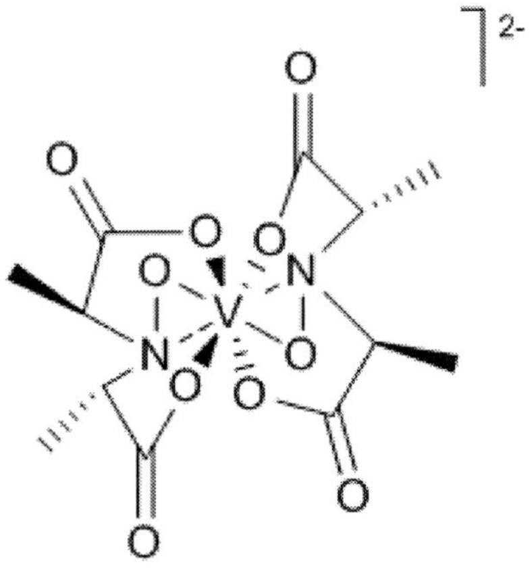

Figure 4. Schematic molecular structure of amavadine

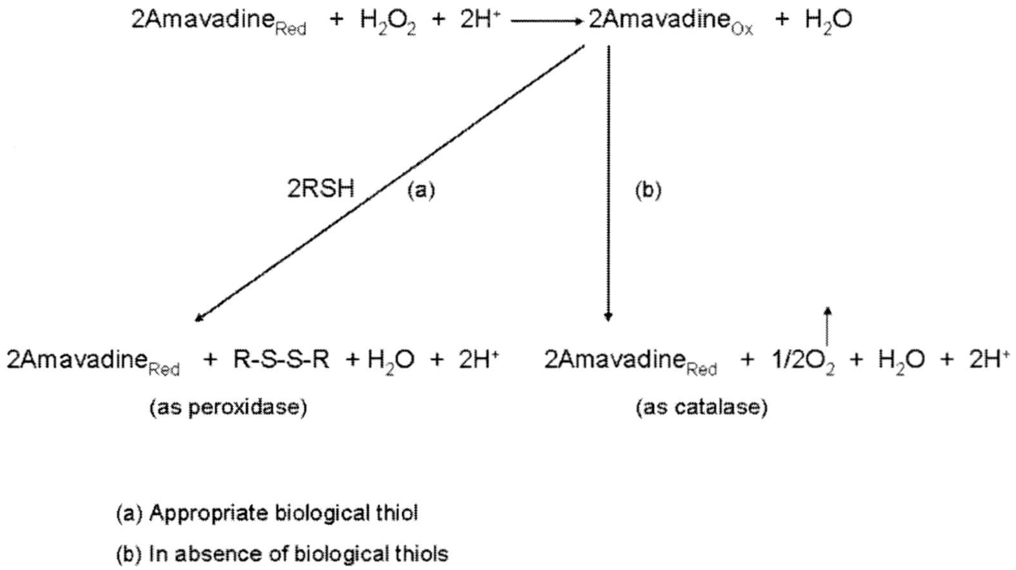

Figure 5. Plausible dual behavior of amavadine (adapted from [13,63]); Amavadine$_{Red}$ is with vanadium in oxidation state four and Amavadine$_{Ox}$ is with vanadium in oxidation state five

Moreover, amavadine exhibits chiral features that can lead to some specific behavior for its mediation in reactions occurring in the fungi or with some biological chiral molecules [13]. On the other hand, amavadine models have been used with success as catalysts for the hydroxylation, oxygenation, peroxidative halogenation and carboxylation of several hydrocarbons, leading to the corresponding alcohols, ketones, organohalides or carboxylic acids. Nevertheless, the experimental conditions used in all of these reactions are not appropriate for their occurrence in fungi, such as the type of solvents [13].

Impacts of Vanadium in Higher Animals

The essentiality of vanadium for humans is not known [74], despite the use of this metal in dietary supplements [75], although for some mammals and other higher animals it is necessary for their metabolic activity at low levels [74]. However, for these organisms the biomolecules containing this metal it have not yet been identified. In spite of vanadium requirement for normal metabolism of some higher animals this contribution for natural cycle of this metal is negligible (placing away the human technological applications that is matter not considered in present chapter). Note that toxic effects of this metal occur for its higher amounts as consequence of its interferences in vanadium non-dependent metabolic performances. Taking into account the values of the oxidation potential in intracellular environments, vanadium is typically present at oxidation state four in biological media, but in some specific sites with higher values of potential is possible its presence at oxidation state five [76].

In case of rats and chickens vanadium deficiency results in reduced growth of their bodies and affected their reproductive activities [77]. In rats its scarcity influence thyroid metabolism, affecting plasma levels of triiodothyronine (a tyrosine derivative) that is a thyroid hormone and it is involved several physiological activities, including the growth [78]. This association suggests a relation between vanadium and an iodine biomolecule in these organisms, which is not unique in life due to occur also in iodoperoxidases vanadium-dependent (see above, Vanadium roles in living organisms), despite it is not identified the pathway between the two chemical elements in higher animals, being an enzyme containing selenium involved in the biosynthesis of active metabolic iodine species [8]. In related to goats, these animals present extra abortion rates and decrease its milk production for vanadium lacking. Additionally, the shortage of this metal in same animals causes shift of levels of several important biomolecules [78].

In humans, vanadium can be harmful, being the effect in organism dependent of metal dose, type of compound, period of exposition and type of contact (breathing, eating, drinking, or by skin). Note that the response of organism is associated with age, gender, diet (and lifestyle, in general), family traits, and state of health. Multiple effects caused by vanadium toxicity are know, such as abdominal pain, nausea, vomiting, diarrhoeas, severe acute renal failure, loss of appetite and pulmonary haemorrhage [79].

Vanadium in Chemical Elements Cycles

Each chemical element of Earth flows among lithosphere, hydrosphere, and atmosphere (if occurs as volatile compounds or as aerosols), and even in biosphere despite not all are essential in the metabolism of organisms, and it should note that some can be pernicious effects in biological systems. An essential chemical element has typically intricate relationships in its overall circulation through both biotic and abiotic compartments of Earth, being their flows based on chemical and physical, and also biochemical procedures. However, the contribution of biosphere is limited due to its mass is much lower than all the other geospheres, but in some cases can provoke localized chief effects and therefore interfere in other reservoirs of the Earth [80].

All living organisms are constituent by some chemical elements and as consequence participate in their recycling mainly for the essential, but in some conditions also for the non-essentials. These flows among organisms occur along the trophic chains (and in symbiotic relationships) and also the life forms and the environments, due to the organisms need regenerate some of their components by external acquisition (*i.e.* from Earth's reservoirs with available material). However, each chemical element is interdependent of others, leading, for example, that its availability is dependent from diverse factors, such as, the pH of the environment and the characteristics of its compounds in media. Low bioavailability of a chemical element affects its occurrence in the organisms and as consequence it is may reduce their survival. A strategy of the organism to avoid this inconvenience is the storage of essential chemical elements in non-toxic conditions. Note that an environment with sudden low availability of one or more essential chemical element can provoke a declining in biodiversity in the ecosystem [81].

The number of elements chemical has been surveyed in their biogeochemicals cycles is partial. Note that in the cases where cycles have been characterized some aspects are not enough detailed for an accurate understanding of their overall cycles. Only two groups have received wide study concerning their global cycle information: (*a*) the main biological components, including the nitrogen, and also for halogens, and (*b*) the toxic chemical elements. A few others have extensive country-scale information but little on a broader scale, such as, vanadium, which has an anthropogenic dominance of its mobilization flows [82].

Vanadium plays a limited role in biology, being more significant in ocean than in land environments (see above, Vanadium roles in living organisms), and it has specific relationships with other chemical elements. Vanadium contribution is more relevant in the cases that its related chemical element is present in, at least, one Earth's compartment, with significant importance, facilitating the flows among geospheres, as happens for some halogens and nitrogen. Halogenated compounds, despite its low levels in Earth, have considerable effects in atmosphere, and nitrogen compounds are very important in atmosphere and in biosphere. Both examples have also implications in ecosystems where they are formed and their derivatives are produced by contribution of vanadium biological activity.

Next subsections are about vanadium relationships with halogen and nitrogen compounds related to their natural cycles.

Vanadium in Halogen (Chlorine, Bromine and Iodine) Cycles

Natural halogenated compounds are almost 5,000 from both abiogenic and biogenic sources [83]. In former group are associated to geothermal processes, such as, volcanic activity and more than a hundred of different compounds has been detected, especially containing chlorine, but the bromine derivatives are also in a significant number [84]. Biological halogenated compounds are mainly from marine environments and occur in wide range of organisms, such as, marine fungi, phytoplankton, macroalgae, marine invertebrates (sponges, molluscs, tunicates, echinoderms and so on), terrestrial plants, fungi and lichen, bacteria, insects, higher animals, including humans. Chlorine halogenated metabolites are the most abundant in variety (slight more than half of all), followed by bromine (together with chlorine are roughly overall known), while iodine and fluorine are less significant. This dissimilar distribution is dependent of abundance, availability and reactivity of halogens [83].

The type of halogenated compounds is diverse and some have different halogens in same molecule [83]. Structural classes range of biological halogenated compounds are varied, from simple phenolic and aliphatic compounds (including, volatile organic halogenated compounds) to more complex structures, such as, oligopeptides [85]. The halometabolites present antibacterial, antifungal, antiviral, anti-inflammatory, antiproliferative, antifouling, antifeeding activity against predators, protection to abiotic stresses, tissue repair, holdfast adhesion, against reactive species generated by oxidative processes, signaling molecules or causing cell wall modifications, and in its adhesion properties [86-88]. In general, the activity of metabolites may permit additional advantages to living organisms in their environments facilitating their adaptation to plausible hostile milieu. These contribute can give support to evolution and biodiversity despite these molecules corresponding typically to a very low percentage in the organisms [89].

In biosynthesis of halogenated compounds are implicated some types of halogenation enzymes. Haloperoxidases are one of the classes of these biological molecules that are less specific than halogenases, but are known other enzymes with halogenation capacities [85]. In group of haloperoxidases, vanadium-dependent enzymes are only a sub-group, which are linked to the halogen cycles (except fluorine, whose anion can not be substrate of these vanadium-dependent enzymes, see above Vanadium roles in living organisms).

Examples of halogenated compounds and their relationships with vanadium-haloperoxidases will be present in next paragraphs.

Iodine is poorly abundant in natural environments, being its main source the oceans, but this non-metal is spread in several kingdoms of life. Terrestrial environments release also volatile iodo-organic compounds from rice paddies [90] and soils [91], suggesting that bacteria from these milieus have enzymes, such as vanadium-haloperoxidases with capacity to produce these compounds. It should be noted that the accumulation of radioactive iodine isotopes and its emission as volatile compounds can affect the ecosystems [92].

Several seaweeds contain iodo- and bromoperoxidases and directly or indirectly take part in the production of the huge amounts of brominated, iodinated compounds, including volatile short-lived organo-iodines, such as methyl iodide, and the formation of I_2. An important group containing vanadium-haloperoxidases are from brown algae that are a major component of littoral and sublittoral zones in temperate and subtropical ecosystems. Some use bromine as substrate, but some orders of this group of seaweeds, such as Laminariales, accumulate and use iodide [88], being to *Laminaria digitata* (Hudson) J.V.Lamouroux 1813 roughly 30,000 more concentrated relatively seawater comparing its dry weight [92], producing volatile compounds, including I_2, which are released by seaweeds at low tide. They have impacts mainly in tropospheric photochemistry due to these compounds are broken down and they produce very reactive iodine radicals that react with ozone present in atmosphere [19,92].

Other iodometabolites are iodotyrosines that occur in various eukaryotic phyla, including algae, in cell-cell communication of plants and in higher organisms as hormones. Vanadium-haloperoxidases are involved in some algae in this metabolic role, but in higher organisms do not used vanadium metallobiomolecules in the biosynthesis of molecules containing iodine [93]. Additionally, ascidians usually accumulate vanadium and have also iodotyrosines, but probably no metabolic relationship between two chemical elements occurs in these organisms [94,95]. However, is possible that vanadium may affect iodine metabolism and thyroid function in rats, but is not involved directly in synthesis of iodometabolites (see also above, Impacts of vanadium in higher animals) [96].

Bromoform ($CHBr_3$) and other bromo-organic volatile compounds are produce by seaweeds and phytoplankton. The amount of bromoform from these sources is probably around 70% of global production, *ca.* 2.2 x 10^{11} g [97]. Volatile halogen compounds have greenhouse effect and react with ozone present in the atmosphere.

These compounds have different residence periods in this geosphere compartment and in some cases persist there for years or decades. Stress conditions must be associated to production and release for atmosphere of these halogenated compounds and some other bromometabolites can be also formed [19]. Vanadium-haloperoxidases probably give a relevant supply for production of these volatile brominated compounds, but some of these are originated from volcanic activity [98].

Vanadium-chloroperoxidases occur in a group of common terrestrial fungi and they can be linked with natural presence of high molecular weight chloro-aromatics in the environment and their enzymatic activity are probably involved in the degradation of plant structures and breakdown of lignin that is the most abundant aromatic substance in Earth. Their association with global cycle of chlorine gives to these enzymes a significant role in the process [99]. On the other hand, the vanadium-chloroperoxidase is associated to synthesis of antibiotics [20], which can contribute to regulation of some biological species in ecosystems.

The synthesis of halo-compounds can be related with oxidative stress that is connected with reactive oxygenated species, such as the H_2O_2 [88], which is a reagent in reactions of haloperoxidases, see eq. 1. Thus, the haloperoxidases and probably also the amavadine (see above, Vanadium roles in living organisms) are related to these species that are mainly present in metabolism of aerobic organisms. Vanadium is probably associated, at least in referred organisms, in regulation of H_2O_2 that to higher levels cause pernicious side effects in biological systems, but apparently has different performances (haloperoxidases versus amavadine) despite some related chemical properties of metal centers of these vanadium metallobiomolecules (see below, Some environmental consequences resulting from vanadium accumulation by some organisms). Curiously, amavadine models are able to catalyze the peroxidative halogenation of some organic substrates [100], regardless of this reaction was achieved in different conditions than by haloperoxidases.

Vanadium in Nitrogen Cycle

Nitrogen belonging to group 15 of periodic table and is a key chemical element for all living organisms, due to be a fundamental constituent of several main biomolecules, such as, nucleic acids and proteins. Its overall biogeochemical cycle links atmospheric, terrestrial, aquatic, and biological reservoirs and is basic to sustain life on Earth, being an essential chemical element that limits primary production in many terrestrial and aquatic ecosystems (less restraining, for example, in tropical forests) and exerts an important control on cycles of many other chemical elements in soils and in oceans [8,101,102]. Nitrogen is the fourth most abundant element in biomass and the atmosphere is the largest reservoir of this chemical element on Earth. However, nitrogen cycle is not stable over history of Earth and it have been affected by oxygenation of atmosphere, as consequence of Great Oxidation Event, leading to establishment of an oxygen-rich atmosphere and causing a different speciation of nitrogen compounds in all geopheres [103].

In its main natural reservoir nitrogen occurs predominantly as an inert compound, N_2, and the vast majority of living organisms are incapable to directly employ of nitrogen from atmosphere. This non-metal as a broad range of oxidation states varying from -3 to 5. In biosphere, it is involved in a cascade of oxidative and reductive reactions. Almost all steps involving reactions of nitrogen in living organisms necessitate some trace metal redox-sensitive in active center. Iron is the main contributor for this role, copper and molybdenum participate also in some enzymes, as well as, vanadium and nickel with less examples. The requirement of these transition metals is a consequence of their redox behavior and their capacity to form enough strong bonds with small nitrogen molecules. However, the bioavailability of these metals, and types of nitrogen compounds, has been altered in the course of time on Earth as consequence of changing of redox conditions causing diverse features on nitrogen cycle during the biological history our planet [8,101,103]. Great number of metallobiomolecules is employed currently in reactions using nitrogen compounds as substrates, which involve the interaction of many organisms. Several types of prokaryotic organisms have a main role in conversion of nitrogen compounds in biosphere, but, such as, fungi and plants gives also a significant contribution, being the reactions both oxidative or reductive [8,101,103].

Vanadium is a minor contributor among the metal transition in modern nitrogen cycle participating in active center of nitrogenases (although in an alternative enzyme) and in vanadium nitrate reductase (despite its occurrence is yet controversial, see above, nitrate reductases). However, the relevance of nitrogen in biomass formation and the significant contribution of vanadium enzyme in uptake of the non-metal from its most relevant reservoir on Earth (atmosphere) give to the metal a considerable participation on modern nitrogen cycle.

The nitrogenases carry out reduction of N_2 (from atmosphere) to NH_3, giving the major contribution of fixed nitrogen to the biogeochemical nitrogen cycle, and are responsible for the cycling of some 10^8 ton of nitrogen per year from the atmosphere to the soil, thereby making it available to other trophic levels due to this chemical element is ubiquitous in living organisms. Typically, the active center of these enzymes contains molybdenum in a cluster with iron and sulfur, but alternative systems are known with vanadium or iron instead molybdenum. Metal bioavailability is a limit for biological nitrogen fixation controlling by this way the ecosystems. Under conditions of metal scarcity, intracellular metals are employed with remarkable efficiency. The occurrence of vanadium in nitrogenases is a consequence unavailability of molybdenum. Note that the average content of molybdenum is a hundred times lower than vanadium in Earth's crust, being necessary a concentration of 25 nM of the former for its uptake, and soils with higher content of sulfur reduce more significantly the availability of the molybdenum than the vanadium. Additionally, vanadium-nitrogenases is much more efficient at lower temperatures than molybdenum enzymes [29,101,104].

Furthermore, vanadium-nitrogenases comprises a specific capacity of CO-reducing, suggesting this characteristic a possible another role of a precursor of this current enzyme as contributor in olden carbon cycles [105]. In view of the importance of CO in some autotrophic organisms, probably more significant in ancient Earth, and its chemical properties that allow the formation of stable complexes with low valence states of the metal in metallobiomolecules, leading to their inhibition [106], it is also possible that vanadium-nitrogenases can be involved in decrease of this pernicious effect in diazotrophic bacteria. Additionally, the feature of vanadium-nitrogenases related to CO may be a result of its shift of role over history of Earth that permits currently a broader spectrum of reactions.

Taking into account the dependence of life by nitrogen and from its main reservoir in Earth (atmosphere), where its chemical species more abundant is chemically inert, the organisms have been developed other strategies to recover this non-metal that is also present in diverse type of chemical species in other geospheres. Some of its inorganic species containing also oxygen resulting from other sources, such as, originated from biological activities. The organisms have several enzymes, which have various transition metals in their active sites, participate in the recycling of the nitrogen oxygenated species, being admitted to nitrite a pivotal role in current nitrogen cycle [107]. Only a vanadium enzyme, a specific nitrate reductase [41,42], is probably involved in this route using nitrate as substrate producing nitrite, see eq. 4, and in case to be proven its specific role (see above, Vanadium roles in living organisms), it is a marginal contribution to overall nitrogen cycle in contrast to vanadium-nitrogenases.

Some Environmental Consequences on Vanadium Availability Due to Its Use As Substrate in Biological Redox Reactions

Vanadium can be substrate for primary electron acceptor in dissimilation or respiration roles due to its occurrence in several oxidation states in potentials allowed by living organisms (see above, Vanadium roles in living organisms). However, in the early contact of an organism with vanadium that does not use this metal in its metabolism, it can reply by changing metal oxidation state to a less toxic chemical species.

Successive adaptations of this living organism may permit further utilization of vanadium (possibly in another chemical form) for more specific biological activities. These roles can permit vanadium participation in specific levels of some ecosystems and can give it an input for evolution. However, the increase of vanadium availability species in ecosystems can provoke pernicious effects in some living organisms not dependent of this metal [8].

On the other hand, vanadium as substrate for primary electron acceptor or respiratory functions can be replaced by several other chemical elements, including non-essential, if its participation is only associated to a chemical reaction based on the values of potential. In the case of vanadate it can be substituted by oxygenated ions from metals or non-metals [44-49].

Some Environmental Consequences Resulting from Vanadium Accumulation by Some Organisms

Organisms with the capacity to accumulate vanadium can survive until the natural resources of this metal in environment are not substantially affected. Great reduction can lead high levels of energy spending to compensate its deficiency that can undermine organism's survival.

Three cases are above-mentioned as vanadium accumulation by organisms, *i.e.*, ascidians, fan-worms and few *Amanita* fungi. The most efficient in terms of higher ratio of accumulation relating to their environments are organisms living in oceans. Their effectiveness for the uptake of this metal is enormous and the lower limit of the metal concentration in seawater that disturbs the organism is unlikely to be reached. Note that for these organisms, if the accumulation of vanadium is associated to defensive strategies, lower levels can affect their survival.

Vanadium relationships with other chemical elements in metabolism are poorly understood for vanadium accumulator organisms.

In particular case of amavadine, as were above-mentioned occurs in few *Amanita* fungi, the relative accumulation can be associated with low efficiency of the vanadium referred biomolecule [63], but is not confirmed this hypothesis. Curiously, each bound hydroxyimino group in amavadine is isoelectronic with the peroxo group formed in haloperoxidases containing vanadium [108], although major differences occur between *Amanita* fungi and terrestrial fungi containing chloroperoxidases.

Final Remarks

The amount of vanadium in biosphere is not significant concerning the terrestrial and marine environments and its uptake by organisms does not cause direct effects in ecosystems, but implications on biogeochemical cycles of the other chemical elements is possible by competition in some biochemical reactions or affecting the availability of other chemical elements. However, the information on biological chemistry of vanadium, its biogeochemical cycle and its influences in the cycles of other chemical elements is yet limited.

Vanadium has an effective presence in biogeochemistry cycles of some chemical elements, but its performance is also dependent of the occurrence of others that can substitute it. The example of molybdenum in nitrogenase is the well known, which are related to abundance and availability of this metal and also from thermodynamic characteristics of the environment. However, other cases are possible including randomly effects, such as, in the case of vanadium as primary electron acceptor due to it can be replaced by several other chemical elements and some are not essential to other organisms that can lead to production of toxic species invasive for some ecosystems. These aspects may contribute to evolution due to the development of specific adaptation strategies in reduction of toxicity.

Vanadium metallobiomolecules can participate as buffer potential regulators in life forms, as protectors against predators by its accumulation in parts of the bodies or involving in synthesis of compounds difficult to be metabolized by other living organisms (such as, the halogenated compounds), are involved in biocatalysis, or give additional protection to some biomolecules.

The majority of cases about vanadium in biology are known for less than three decades (the exceptions are those related to ascidians and *Amanita* fungi), which does not discard the possible existence of other examples. Additionally, are still poorly known its roles in living organisms, issues that have effects in the cycles of other chemical elements in natural environments. Advances in these topics will give more appropriate explanation about multiple relationships between the various chemical elements in Earth and the contribution of biosphere in these processes. However, the imprint of human activity in Earth leaves its mark on the planet, which needs to be taking into account in biogeochemical cycles, due to the increasing consumption of natural resources.

Acknowledgments

This contribution was supported by pluriannual project PEst-OE/QUI/UI0100/2011 funding from Fundação para a Ciência e a Tecnologia (FCT), Portugal.

References

[1] Miyauchi, A.; Okabe, T. H. *J Jpn Inst Met* 2010 *74* 701-711.
[2] *In Encyclopedia of Inorganic Chemistry*, Editor, King, R. B.; *2nd Ed.*; John Wiley & Sons: *Chichester*, U.K., 2005 (vanadium issues in volume IX).
[3] Rehder, D. *Org Biomol Chem* 2008 *6* 957-964.

[4] Och, L. M.; Shields-Zhou, G. A. *Earth-Sci Rev* 2012 *110* 26-57.
[5] Wang, D.; Wilhelmy, S. A. S. *Mar Chem* 2009 *117* 52-58.
[6] Emerson, S. R.; Huested S. S. *Mar Chem* 1991 *34* 177-196.
[7] Bleackley, M. R.; MacGillivray, R. T. A. *Biometals* 2011 *24* 785-809.
[8] Fraústo da Silva, J. J. R.; da Silva, J. A. L. *Os Elementos Químicos e a Vida*, IST Press: Lisboa, 2011.
[9] Mukherjee, B.; Patra, B.; Mahapatra, S.; Banerjee, P.; Tiwari, A.; Chatterjee, M. *Toxicol Lett* 2004 *150* 135-143.
[10] Bertrand, D. In *Survey of Comtemporany Knowledge of Biogeochemistry*; Bulletin of the American Museum of Natural History, New York, 1950; Vol. 94, pp. 403-456.
[11] Ye, J. R.; Qing, H. R.; Bend, S. L.; Gu, H. R. *AAPG Bull* 2007 *91* 1167-1188.
[12] Jeong, G. Y.; *Miner Depos* 2006 *41* 469-481.
[13] da Silva, J. A. L.; Fraústo da Silva, João J. R.; Pombeiro, A. J. L. In *Vanadium: Biochemical and Molecular Biological Approaches*; Editor, Michibata, H.; Springer: 2012; pp. 35-49.
[14] Hope, B. K. *Sci Total Environ* 1994 *41* 1-10.
[15] Michibata, H.; Ueli, T. In *Vanadium: Biochemical and Molecular Biological Approaches*; Editor, Michibata, H.; Springer: 2012; pp. 51-71.
[16] Fattorini, D.; Regoli, F. In *Vanadium: Biochemical and Molecular Biological Approaches*; Editor, Michibata, H.; Springer: 2012; pp. 73-92.
[17] Garg, N.; Chandel, S. *Agron Sustain Dev* 2010 *30* 581-599.
[18] Bonfante, P., & Genre, A. (2010) Mechanisms underlying beneficial plant-fungus interactions in mycorrhizal symbiosis. *Nat Commun 1.*
[19] Wever, R. In *Vanadium: Biochemical and Molecular Biological Approaches*; Editor, Michibata, H.; Springer: 2012; pp. 95-125.
[20] Bernhardt, P.; Okino, T.; Winter, J. M.; Miyanaga, A.; Moore, B. S. *J Am Chem Soc* 2011 *133* 4268-4270.
[21] Blasiak, L. C.; Drennan, C. L. *Accounts Chem Res* 2009 *42* 147-155.
[22] Leak, D. J.; Sheldon, R. A.; Woodley, J. M.; Adlercreutz, P. *Biocatal Biotransfor* 2009 *27* 1-26.
[23] Messerschmidt, A.; Prade, L.; Wever, R. *Biol Chem* 1997 *378* 309-315.
[24] Weyand, M.; Hecht, H. J.; Kiess, M.; Liaud, M. F.; Vilter, H.; Schomburg, D. *J Mol Biol* 1999 *293* 595-611.
[25] Colin, C.; Leblanc, C.; Michel, G.; Wagner, E.; Leize-Wagner, E.; van Dorsselaer, A.; Potin, P. *J Biol Inorg Chem* 2005 ***10*** 156–166.
[26] Almeida, M.; Humanes, M.; Melo, R.; Silva, J. A.; Fraústo da Silva, J. J. R.; Vilter, H.; Wever, R. *Phytochemistry* 1998 *48* 229-239.
[27] Hasan, Z.; Renirie, R.; Kerkman, R.; Ruijssenaars, H. J.; Hartog, A. F.; Wever, R. *J. Biol Chem* 2006 *281* 9738-9744.
[28] ten Brink, H. B.; Tuynman, A.; Dekker, H. L.; Hemrika, W.; Izumi, Y.; Oshiro, T.; Schoemaker, H. E.; Wever, R. *Inorg Chem* 1998 *37* 6780-6784.
[29] Eady, R. R. *Coord Chem Rev* 2003 *237* 23–30.
[30] Oda, Y.; Samanta, S. K.; Rey, F. E.; Wu, L. Y.; Liu, X. D.; Yan, T. F.; Zhou, J. Z.; Harwood, C. S. *J Bacteriol* 2005 *187* 7784-7794.
[31] Fisher, K.; Dilworth, M. J.; Newton, W. E. *Biochemistry* 2006 *45* 4190–4198.
[32] Lee, C. C.; Hu, Y.; Ribbe, M. W. *Proc Natl Acad Sci USA* 2009 *106* 9209–9214.

[33] Lovell, T.; Torres, R. A.; Han, W. G.; Liu, T.; Case, D. A.; Noodleman, L. *Inorg Chem* 2002 *41* 5744–5753.
[34] Wichard, T.; Bellenger, J. P.; Morel, F. M. M.; Kraepiel, A. M. L. *Environ Sci Technol* 2009 *43* 7218-7224.
[35] Fisher, K.; Lowe, D. J.; Petersen, J. *Chem Commun* 2006 2807-2809.
[36] Gupta, K. J.; Fernie, A. R.; Kaiser, W. M.; van Dongen, J. T. *Trends Plant Sci* 2011 *16* 160-168.
[37] Mendel, R. R.; *Biofactors* 2009 *35* 429-434.
[38] Antipov, A. N.; Sorokin, D. Y.; L'Vov, N. P.; Kuenen, J. G. *Biochem J* 2003 *369* 185–189.
[39] Filimonenkov, A. A.; Zvyagilskaya, R. A.; Tikhonova, T. V.; Popov, V. O. *Biochem-Moscow* 2010 *75* 744-751.
[40] Antipov, A. N.; Lyalikova, N. L.; L'Vov, N. P. *IUBMB Life* 2000 *49* 137–141.
[41] Antipov, A. N.; Lyalikova, N. L.; Khijniak, T. V., L'Vov, N. P. *FEBS Lett* 1998 *441* 257–260.
[42] Antipov, A. N.; Lyalikova, N. L., Khijniak, T. V., L'Vov, N. P. *IUBMB Life* 2000 *50* 39–42.
[43] Al-Samarrae, K. W.; Al-Ahmed, H. I.; Abdulatif, W. *Journal of Biotechnology Research Center* 2010 *4* 3-10.
[44] Carpentier, W.; De Smet, L.; van Beeumen, J.; Brigé, A. *J Bacteriol* 2005 *187* 3293–3301.
[45] Carpentier, W.; Sandra, K.; De Smet, I.; Brigé, A.; De Smet, L.; van Beeeumen, J. *Appl Environ Microbiol* 2003 *69* 3636–3639.
[46] Yurkova, N. A.; Saveleva, N. D.; Lyalikova, N. N. *Microbiology* 1993 *62* 367-370.
[47] Moritani, C.; Sakai, Y.; Tsuda, M.; Kanazawa, H.; Tschuchiya, T. *Chem Pharm Bull* 1990 *38* 164–167.
[48] Ortiz-Bernad, I.; Anderson, R. T., Vrionis, H. A.; Lovley, D. R. *Appl Environ Microbiol* 2004 70 3091–3095.
[49] Csotonyi, J. T.; Stackebrandt, E.; Yurkov, V. *Appl Environ Microbiol* 2006 *72* 4950–4956.
[50] Odate, S.; Pawlik, J. R. *J Chem Ecol* 2007 33 643-654.
[51] Shenkar, N.; & Swalla, B. J. (2011) Global Diversity of Ascidiacea. *PLoS One 6*.
[52] Michibata, H.; Iwata, Y.; Hirata, J. *J Exp Zool* 1991 *257* 306-313.
[53] Hirata, J.; Michibata, H. *J Exp Zool* 1991 *257* 160-165.
[54] Macara, I. G.; McLeod, G. C.; Kustin K. *Comp Biochem Phys* 1979 *62* 821-826.
[55] Ueki, T.; Michibata, H. *Coord Chem Rev* 2011 *255* 2249-2257.
[56] Yamaguchi, N.; Amakawa, Y.; Yamada, H.; Ueki, T.; Michibata H. *Zool Sci* 2006 *23* 909-915.
[57] Young, C. M.; Vazquez, E. *Invertebr Biol* 1997 *116* 262-276.
[58] Ishii, T.; Nakai, I.; Numako, C.; Okoshi, K.; Otake, T. *Naturwissenschaften* 1993 *80* 268-270.
[59] Fattorini, D.; Notti, A.; Nigro, M.; Regoli, F. *Environ Sci Pollut Res* 2010 *17* 220-228.
[60] Glasby, C. J.; Timm, T. *Hydrobiologia* 2008 *595* 107-115.
[61] Meisch, H.-U.; Reinle, W.; Schmitt, J. A. *Z Naturforsch* 1979 *66* 620-621.
[62] Koch, E.; Kneifel, H.; Bayer, E. *Z Naturforsch* 1987 *42c* 873-878.

[63] Matoso, C. M. M.; Pombeiro, A. J. L.; Fraústo da Silva, J. J. R.; da Silva, M. F. C. G.; da Silva, J. A. L.; Baptista-Ferreira, J. L.; Pinho-Almeida, F. In *Vanadium Compounds – Chemistry, Biochemistry and Therapeutic Applications*; Editor, Tracey, A. S.; Crans, D. C.; American Chemical Society Symposium Series N° 711; American Chemical Society, Washington, 1998; pp. 241-247.

[64] Geml, J.; Tulloss, R. E.; Laursen, G. A.; Sazanova, N. A.; Taylor, D. L. *Mol Phylogenet Evol* 2008 *48* 694-701.

[65] Michelot, D.; Melendez-Howell, L. M. *Mycol Res* 2003 *107 131-146.*

[66] Armstrong, E. M.; Beddoes, R. L,.; Calviou, L. J.; Charnock, J. M.; Collison, D.; Ertok, N.; Naismith, J. H.; Garner, C. D. *J Am Chem Soc* 1993 *115* 807-808.

[67] Berry, R. E.; Armstrong, E. M.; Beddoes, R. L.; Collison, D.; Ertok, S. N.; Helliwell, M.; Garner, C. D. *Angew Chem Int Engl* 1999 *38* 795-797.

[68] Fraústo da Silva, J. J. R.; Guedes da Silva, M. F. C.; Silva, J. A. L.; Pombeiro, A. J. L. In *Molecular electrochemistry of inorganic, bioinorganic and organometallic compounds*, Editor, Pombeiro, A. J. L.; McCleverty, J. A. Kluwer Academic Publishers, The Netherlands, 1993; pp.411-415.

[69] Guedes da Silva, M. F. C.; Silva, J. A. L.; Fraústo da Silva, J. J. R.; Pombeiro, A. J. L.; Amatore, C., Verpeaux, J.-N. *J Am Chem Soc* 1996 *118* 7568-7573.

[70] Nawi, M. A.; Riechel, T. L. *Inorg Chim Acta* 1987 *136* 33-39.

[71] Thackrey, R. D.; Riechel, T. L. *J Electroanal Chem* 1988 *245* 131-143.

[72] Carrondo, M. A. A. F. de C. T.; Duarte, M. T. L. S.; Costa Pessoa, J.; Silva, J. A. L.; Fraústo da Silva, J. J. R.; Vaz, M. C. T. A.; Vilas-Boas, L. F. *J Chem Soc, Chem Commun* 1988 1158-1159.

[73] Silva, J. A. L. *Ph'D Thesis, IST.* Lisboa, Portugal, 1992.

[74] Mukherjee, B.; Patra, B.; Mahapatra, S.; Banerjee, P.; Tiwari, A.; Chatterjee, M. *Toxicol Lett* 2004 *150* 135-143.

[75] Repinc, U.; Benedik, L. *Acta Chim Slov* 2004 *51* 59-65.

[76] Sakurai, H.; Fujii, K.; Watanabe, H.; Tamura, H. *Biochem Biophys Res Commun* 1995 ***214*** 1095-1101.

[77] Schwarz, K.; Milne D. B. *Science* 1971 *174* 426-428.

[78] Badmaev, V.; Prakash S.; Majeed, M. *J Altern Complement Med* 1999 ***5*** 273–291.

[79] Valko, M.; Morris, H.; Cronin, M. T. D. *Curr Med Chem* 2005 *12* 1161-1208.

[80] Levin, S. A. *Ecosystems* 1998 *1* 431–436.

[81] Williams, R. P. J.; Fraústo da Silva, J. J. R.; *The Chemistry of Evolution – The Development of our Ecosystem*, Elsevier: Amsterdam, 2006.

[82] Klee, R. J.; Graedel, T. E. *Annu Rev Env Resour* 2004 *29* 69-107.

[83] Herrera-Rodriguez, L. N.; Khan, F.; Robins, K. T.; Meyer, H. P. *Chim Oggi-Chem Today* 2011 *29* 31-33.

[84] Gribble, G. W. *Chemosphere* 2003 *52* 289-297.

[85] Wagner, C.; El Omari, M.; Konig, G. M. *J Nat Prod* 2009 *72* 540-553.

[86] Blunt, J.W.; Copp, B.R.; Hu, W.P.; Munro, M.H.G.; Northcote, P.T.; Prinsep, M.R. *Nat Prod Rep* 2009 *26* 170–244.

[87] Winter, J. M.; Moore, B. S. *J Biol Chem* 2009 *284* 18577-18581.

[88] La Barre, S.; Potin, P.; Leblanc, C.; Delage, L. *Mar Drugs* 2010 *8* 988-1010.

[89] Cabrita, M. T.; Vale, C.; Rauter, A. P. *Mar Drugs* 2010 *8* 2301-2317.

[90] Redeker, K. R.; Wang, N.; Low, J. C.; McMillan, A.; Tyler, S. C.; Cicerone, R. J. *Science* 2000 *290* 966–969.
[91] Amachi, S.; Kasahara, M.; Hanada, S.; Kamagata, Y.; Shinoyama, H.; Fujii, T.; Muramatsu, Y. *Environ Sci Technol* 2003 *37* 3885–3890.
[92] Leblanc, C.; Colin, C.; Cosse, A.; Delage, L.; La Barre, S.; Morin, P.; Fievet, B.; Voiseux, C.; Ambroise, Y.; Verhaeghe, E.; Amouroux, D.; Donard, O.; Tessier, E.; Potin, P. *Biochimie* 2006 *88* 1773-1785.
[93] Crockford, S. J. *Integr Comp Biol* 2009 *49* 155-166.
[94] Barrinton, E. J. W.; Thorpe, A. *Proc R Soc B* 1968 *171* 91-109.
[95] Webb, D. A. *J Exp Biol* 1939 *16* 499-523.
[96] Uthus, E. O.; Nielsen, F. H. *Magnesium Trace Elem* 1990 *9* 219-226.
[97] Carpenter, L. J.; Liss, P. S. *J Geophys Res* 2000 *105* 20539-20547.
[98] Frische, M.; Garofalo, K.; Hansteen, T. H.; Borchers, R.; Harnisch, J. *Environ Sci Pollut Res* 2006 *13* 406-413.
[99] Ortiz-Bermúdez, P.; Hirth, K. C.; Srebotnik, E.; Hammel, K. E. *Proc Natl Acad Sci USA* 2007 104 3895-3900.
[100] Reis, P. M.; Silva, J. A. L.; da Silva J. J. R. F.; Pombeiro, A. J. L. *Chem Commun* 2000 1845-1856.
[101] [Jaffe, D. A.; Weiss-Penzias, P. S. In *Encyclopedia of Atmospheric Sciences 2003* 205-213.
[102] Gruber, N. In *Nitrogen in the Marine Environment* Editor, Capone, D. G.; Bronk, D. A.; Mulholland, M. R.; Carpenter, E. J. 2nd Ed.; Elsevier: 2008; pp. 1-50.
[103] Godfrey, L. V.; Glass, J. B. In *Methods in Enzymology: Research on Nitrification and Related Processes*; Editor, Klotz, M. G.; Book Series: Methods in Enzymology 2011; Vol. 486 pp.483-506.
[104] Miller, R. W.; Eady R. R. *Biochem J* 1988 *256* 429-432.
[105] Hu, Y. L.; Lee, C. C.; Ribbe, M. W. *Science* 2011 *333* 753-755.
[106] Ragsdale, S. W. *Crit Rev Biochem Mol Biol* 2004 *39* 165-195.
[107] Vlaeminck, S. E.; Hay, A. G.; Maignien, L.; Verstraete, W. *Environ Microbiol* 2011 *13* 283-295.
[108] Reis, P. M.; Silva, J. A. L.; Palavra, A. F.; da Silva, J. J. R. F.; Kitamura, T.; Fujiwara, Y.; Pombeiro, A. J. L. *Angew Chem Int Engl* 2003 *42* 821-823.

In: Vanadium
Editors: V. N. Baranova and A. V. Fortunatov

ISBN 978-1-62081-600-4

Chapter 2

THE ROLE OF VANADIUM IN MICROALLOYED STEELS: AUSTENITE GRAIN REFINEMENT, INTRAGRANULAR NUCLEATION OF FERRITE AND MECHANICAL PROPERTIES

Sebastián F. Medina[1], Manuel Gómez[1], Alberto Quispe[2] and Lucía Rancel[1]

[1]National Centre for Metallurgical Research (CENIM-CSIC), Madrid, Spain

[2]National University Jorge Basadre, University City, Tacna, Peru

ABSTRACT

The most important elements in microalloyed steels are niobium and vanadium. Nb is used in low carbon steels known as HSLA (High Strength and Low Alloy) steels. V is used in HSLA steels as well as in medium and high-carbon steels. Compared to Nb, the advantage of V is the very low solubility temperature of carbides, so the nanoprecipitates are usually vanadium nitrides. However, Nb cannot be used in medium and high-carbon microalloyed steels because the solubility temperature of niobium carbides and carbonitrides is very high and these precipitates do not dissolve at reheating temperatures prior to thermomechanical treatment. In this work, an analysis of V nanoprecipitates in thermomechanically-processed steels is carried out by means of scanning and transmission electron microscopy. It is shown that these particles have a mean size of a few nanometers (< 30 nm) due to their low solubility temperature. In consequence, and according to Orowan equation, VN precipitates cause a more remarkable increase in yield strength than NbCN precipitates. Another benefit of V microalloyed steels introduced in this work is the intragranular nucleation of ferrite and bainite in austenite grains existing before phase transformation. VN particles have low interfacial energy with respect to ferrite but relatively high interfacial energy with respect to austenite for the (001)VN boundary compared with other particles or inclusions. Such advantages promote the intragranular nucleation of ferrite and bainite. This work shows that V is an essential alloying element in high strength (> 1000 MPa) microalloyed steels with a ferrite-pearlite microstructure, as well as in bainitic steels. In low carbon steels, with ultimate strength near 600-700 MPa, Nb is also very interesting as a microalloying element. It is used in lower amounts, due to its higher solubility temperature and consequently its higher no-recrystallization temperature, which allows a more severe strengthening of austenite and a higher increase in dislocation density. Nevertheless, the intragranular nucleation of ferrite and bainite in V microalloyed steels make this element the best choice, provided that optimal additions are applied. Results about the static recrystallization of austenite during hot rolling and the recrystallization-precipitation-time-temperature (RPTT) diagrams will be also presented in this work. RPTT diagrams display interesting results about the minimum time for precipitation, the recrystallized volume fraction that favors precipitation and the optimal rolling conditions to obtain a very fine ferritic grain size. On the other hand, the increase in yield strength in high strength steels due to the presence of VN precipitates and the improvement in toughness in bainitic steels will be shown. An optimal V/N/Mn ratio will be established. The amount of V used in microalloyed steels is usually between 0.10 and 0.20 %, which means there is an important industrial market for this element.

INTRODUCTION

The type and amount of microalloying elements have an important effect on the shape and nature of precipitates. At an equal level of microalloying, the precipitates are soluble in austenite as follows [1,2]: $Nb(C,N)<VN<VC$. Microalloying elements dissolved in austenite have a considerable effect on grain growth, the progress of recrystallization, and on phase transformation. However, the main reason for the presence of these elements in microalloyed steels lies in their precipitation, and particularly in the interaction between particles of these elements (Ti, Nb, V) and interstitials (C, N) with austenite grain boundaries in motion. Whereas solutes hinder the advance of grain boundaries by means of a friction effect originated by the difference in their atomic radius compared to iron, precipitates cause a decrease in the effective grain boundary area and thus in the associated surface energy. The latter leads to an obstruction of grain boundary motion or a pinning effect, which is much stronger than the solute drag.

Nitrides and carbides of typical microalloying elements (Nb,V) have an f.c.c crystallographic structure. These compounds, especially in the case of the smallest particles, frequently form precipitates, which are semi-coherent with the (f c c) austenitic matrix. Their lattice parameter is slightly higher than that of the austenite [3].

The size and volume fraction of precipitates play an important role on pinning forces, both for austenite grain growth control at any reheating temperature and for inhibition of austenite static recrystallization during hot rolling. The latter causes a strengthening of austenite and an increase in dislocation density that in turn multiplies the number of potential nucleation sites for ferrite [4-8].

The no-recrystallization temperature (T_{nr}) represents the start of inhibition of the static recrystallization of austenite during hot rolling. The most common method for determining T_{nr} consists of simulating successive rolling passes and then graphically representing the Mean Flow Stress (MFS) versus the inverse of the absolute temperature for each of the simulated passes. The inhibition of recrystallization indicated by T_{nr} appears as a rise in the slope of the MFS [9-13]. This method also offers the possibility of obtaining the A_{r3} phase transformation temperature when austenite is being deformed in conditions similar to hot rolling and the stress accumulated ($\Delta\sigma$) in the austenite between T_{nr} and A_{r3} [14,15]. The accumulated stress in austenite is a direct and precise measurement of the state of austenite as a function of the temperature and can be directly measured without the need to measure the size and lengthening of the austenite grain [16].

There are basically three types of heterogeneous nucleation sites for ferrite in austenite, namely grain and subgrain boundaries, the latter as stacks of dislocations or polygonization. Precipitates or inclusions VC or VN have a low interfacial energy in relation to ferrite but a relatively high interfacial energy in relation to austenite for the (001)V(C,N) boundary compared to MnS inclusions. MnS type inclusions produce a local Mn impoverishment that increases the A_{r3} value, favoring the nucleation of ferrite. However, the advantages of VC and VN over MnS in the balance of interphase boundary energy presumably promote the intragranular ferrite transformation for complex precipitates [17]. In the hot working of microalloyed steels, all three of the aforementioned nucleation sites can coexist and act simultaneously on ferrite grain refinement [18].

The fine strengthening precipitates are usually formed during hot deformation and austenite–ferrite transformation or austenite-bainite transformation, and results in the formation of small microalloy nitrides/carbides. These nanoparticles are very hard and extensive dislocation bowing is commonly seen. The development of a satisfactory constitutive equation to describe the quantitative effects of dispersed second phase nanoparticles, i.e. the Ashby-Orowan equation, allows the forecast of increments in yield strength of microalloyed steel in terms of the microstructural features [3,19,20].

The present work deals with V-microalloyed steels and several aspects will be seen: dynamic recrystallization, static recrystallization, thermomechanical treatments, ferrite grain refinement, intragranular nucleation of ferrite, precipitate size, and influence of precipitates (VCN) on yield strength, among others.

Dynamic Recrystallization and Flow Curves

When a low alloy steel is strained at relatively high rates in the austenite phase, its flow curves present the typical regions of hot deformation: a hardening region characterized by an increasing dislocation density, followed by a dynamic recovery characterized by a polygonized austenite, and finally dynamic recrystallization where the build-up of dislocations leads to the nucleation and growth of recrystallized grains during the deformation until the steady state is reached where the recrystallization is total and continuous. At first the stress increases until it reaches a maximum (σ_p), where the nucleation of some new grains has already begun, and then descends as recrystallization progresses until it reaches a constant value (σ_{ss}) where recrystallization is 100% and continuous. The strain at which dynamic recrystallization starts has been the object of several studies, and almost all coincide in that it begins at a strain slightly below peak strain (ε_p) [21-24].

Several models have been proposed to calculate the dynamically recrystallized fraction (DRX), though they are only applicable to a certain type of steel family, normally the C-Mn steels [25-27]. It is generally admitted that DRX follows Avrami's law and the most common expression to quantitatively interpret this phenomenon has the following form:

$$X_d = 1 - \left[-k \left(\frac{\varepsilon - a\varepsilon_p}{\varepsilon_p} \right)^{m'} \right] \tag{1}$$

$$\varepsilon > a\varepsilon_p$$

where k and m´ are generally constants, with values between: $0 < k < 1$; $1 < m' < 2$, and ε_p is the peak strain. Some authors have used another equation, though still based on Avrami's law, with a form similar to the equation used for the statically recrystallized fraction [28].

A model was established to calculate the flow curve as a result of subtracting from one expression, which predicts the stress in the hardening region plus the dynamic recovery region, a second expression, which predicts the softening effect of dynamic recrystallization, in other words the model predicts the evolution of the microstructure during deformation. The modelling of dynamic recrystallization is of great importance in the prediction of microstructures in hot rolled steel as it is not normally advisable to reach strains above the strain at which recrystallization begins ($a\varepsilon_p$) as this may give rise to heterogenous structures, as occurs in the conventional strip mill. However, in some types of rolling the aim is to obtain finer structures, as is the case of the newly designed controlled rolling procedures, which involve larger strains and provide significant proportions of dynamic recrystallization that in turn leads to finer austenite grains [29-33]. Medina et al. [34-37] found the following expressions for the parameters k and m´ in equation (1), respectively:

$$k = 0.5974 \exp\left[1.2333 \left(\frac{Z}{A} \right)^{-0.5} \right] \tag{2}$$

$$m' = 1.0901 \exp\left[0.0264 \left(\frac{Z}{A} \right)^{0.5} \right] \tag{3}$$

In V-microalloyed steels, the expression for the peak strain ε_p is [35]:

$$\varepsilon_p = 0.0988 D_0^{0.195} \left(\frac{Z}{A} \right)^{0.166} \tag{4}$$

where, Z is Zener-Hollomon parameter:

$$Z = \dot{\varepsilon} \exp \frac{Q_d}{RT} \tag{5}$$

where is the strain rate and Q_d is the activation energy for austenite hot deformation, given by [34]:

$$Q_d(J.mol^{-1}) = 267000 - 2535.5[C] + 1010[Mn] + 33620.8[Si] + \\ + 70729.9[Nb]^{0.565} + 31673.5[V] + 93680.5[Ti]^{0.5919} \tag{6}$$

where each square bracket indicates the percentage by weight of the element indicated.

The activation energy is sensitive to any change in the chemical composition of the austenite and rises in line with an increase in the alloy contents, except in the case of carbon.

And finally, A is given by:

$$A(s^{-1}) = [12.197 + 65.590(C\%) - 49.052(Nb\%)]\exp(7.076 \cdot 10^{-5} Q_d) \tag{7}$$

Equation (1), together with equations (2)-(7) make it possible to predict the dynamically recrystallized fraction as a function of the strain, temperature, strain rate, initial grain size and chemical composition of a broad selection of steels including V-microalloyed steels.

For the hardening and dynamic recovery region of the flow curve the following expression for the flow stress σ_e has been used [26,38,39]:

$$\sigma_e = B[1 - \exp(-C\varepsilon)]^m \tag{8}$$

Parameters B and C are dependent on the conditions of deformation (temperature and strain rate). Mathematically, B can be expressed as:

$$B = \lim_{\varepsilon \to \infty} \sigma_e \tag{9}$$

being ε the strain.

The work hardening exponent (*m*) of equation (8) is normally taken as being a constant. However, as will be seen later, it is a function of the conditions of deformation and of the chemical composition of the steel. The equation used to predict the reduction of stress for strains greater than ε_p due to the effect of dynamic recrystallization has the following form:

$$\Delta\sigma = B'X_d \tag{10}$$

B' depends on the conditions of deformation and it can be mathematically expressed in the following way:

$$B' = \lim_{\varepsilon \to \infty} \Delta\sigma_e \tag{11}$$

The flow curve is given by the resulting expression $(\sigma_e - \Delta\sigma)$, a function of the deformation and it contains eight parameters (*B, C, m, B', k, m', a* and σ_p) whose expressions are as follows [36]:

$$B(MPa) = \left(9.5326 + 0.6196 \ln \frac{Z}{A}\right)^2 \tag{12}$$

$$C = 3.9202\left(\frac{Z}{A}\right)^{0.0592} \tag{13}$$

$$m = 0.3449 \exp\left(0.0139\sqrt{\frac{Z}{A}}\right) \tag{14}$$

$$B'(MPa)) = 26.031\left(\frac{Z}{A}\right)^{0.1351} \tag{15}$$

a=0. 95

When the above expressions are introduced in the expression $(\sigma_e - \Delta\sigma)$ they predict the flow curves with good accuracy, whatever the temperature, strain rate and chemical composition of the steel, for all the low alloy and V-microalloyed steels. To illustrate the verification of the model, Figure 1 shows the experimental end predicted curves for a V-microalloyed steel.

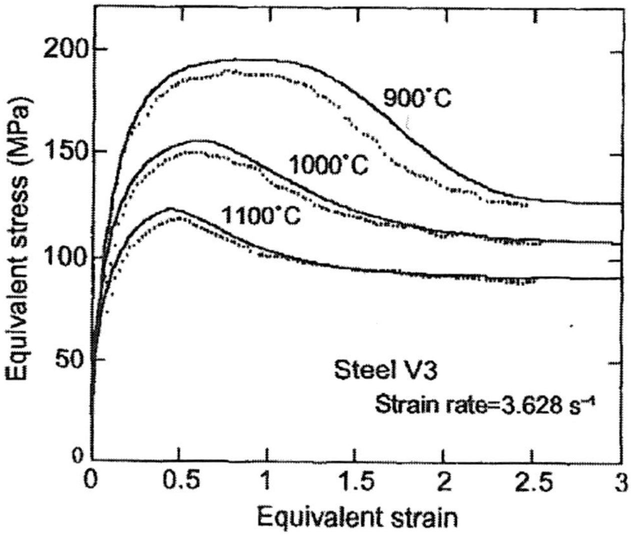

Figure 1. Experimental (- - -) and predicted ( ) flow curves for V-microalloyed steel (C=0.11; Si=0.24; Mn=1; V=0.093; N=0.0144, wt%) at 900°C, 1000°C and 1100°C and rate strain of 3.63 s^{-1}.

Finally, the recrystallized austenite grain size has been determined as a function of Z/A, it being seen that this follows the power law mentioned above, whose regression line obeys the following expression [36]:

$$D_d = 38.26\left(\frac{Z}{A}\right)^{-0.08} \tag{16}$$

The dynamically recrystallized grain size, in accordance with equation (16), gives values which coincide approximately with Sellars' model [24, 39] when the *Z* parameter is high, and with the model developed by Namba *et al* [40], when the Z parameter is relatively low.

STATIC RECRYSTALLIZATION

Though the static recrystallization of austenite started to be studied in the 1970's, the first expressions quantifying the influence of the different variables that intervene in hot deformation (strain, strain rate, temperature and grain size) on static recrystallization kinetics appeared in 1980. The model published by Sellars [24] supposed an important reference for better understanding the role of these variables, and opened up the possibility of modelling the phenomenon of static recrystallization and predicting the evolution of microstructures in hot rolling. Most models have been published since 1990 and a large number of them have been based on expressions similar to that of Sellars, changing the quantitative influence of the different variables from one model to another.

Most of the published models for predicting the static recrystallization of austenite [41-47] have not taken into account the influence of the content of each alloying element, and only some take into account the influence of certain elements. In general, the authors establish a single model for C-Mn type steels, irrespective of their chemical composition, and another for Nb microalloyed steels, irrespective of the Nb content. However, alloying elements certainly exert an influence on static recrystallization, and though the first studies undertaken in this respect only dealt with their qualitative influence [48,49], Medina et al published [50-53] a model that established the quantitative influence of some alloying elements (Si, Mn, Mo, V, Nb, Ti). This influence was quantified by means of an expression for activation energy (Q), in a way similar to dynamic recrystallization. This is because of the dual character of a nucleation and growth reaction. The rate at which a metal recrystallizes depends on the rate at which nuclei form, and on the rate at which they grow. These two rates also determine the final grain size of a recrystallized metal.

The variation in activation energy according to the chemical composition showed that solute atoms interact with the grain boundaries. If the size of a foreign atom and the parent crystal are different, then there will be an elastic stress field introduced into the lattice by each foreign atom. The elements that most distort the lattice structure of austenite have the largest effect on grain boundary migration. The greater the atomic volume of the substitutional elements, the greater the distortion [54].

The static recrystallization kinetics of austenite can be described by an Avrami equation in the following way [24,26]:

$$X_a = 1 - \exp\left[-0.693\left(\frac{t}{t_{0.5}}\right)^n\right] \tag{17}$$

where X_a is the fraction of the recrystallized volume and $t_{0.5}$ is the time corresponding to half of the recrystallized volume, which depends practically on all the variables that intervene in hot deformation and whose most general expression follows a law of the type:

$$t_{0.5} = A\varepsilon^{p}\dot{\varepsilon}^{q}D^{s}\exp\frac{Q}{RT} \tag{18}$$

where ε is the strain, $\dot{\varepsilon}$ the strain rate, D the austenite grain size, Q the activation energy, T the absolute temperature, R=8 3145 $Jmol^{-1}K^{-1}$, and p, q and s are parameters. Thus, the construction of the model will be based on the determination of the magnitudes: *A, Q, p, q, s* and *n*.

Figure 2 shows X_a for one of the V microalloyed steels, where a plateau can be seen to form after a certain temperature. This plateau is a consequence of strain-induced precipitation. The curves corresponding to temperatures where the microalloying elements are still in solution have the typical sigmoidal shape of Avrami's law. The plateau indicates that the recrystallized fraction is temporarily inhibited until it starts to progress again, also following Avrami's law.

The start and end of the plateau have been identified with the start and end of induced precipitation. However, while the start of the plateau seems to coincide with good exactness with the start of induced precipitation, the end of the plateau could coincide with an important growth in the average size of precipitates, which become incapable of inhibiting recrystallization, or that, despite the fact that the average size has grown, there is still a certain volume of small size precipitates uniformly distributed in the austenite matrix [55,56]. Though the two definitions might at first sight seem different, in fact they mean practically the same thing.

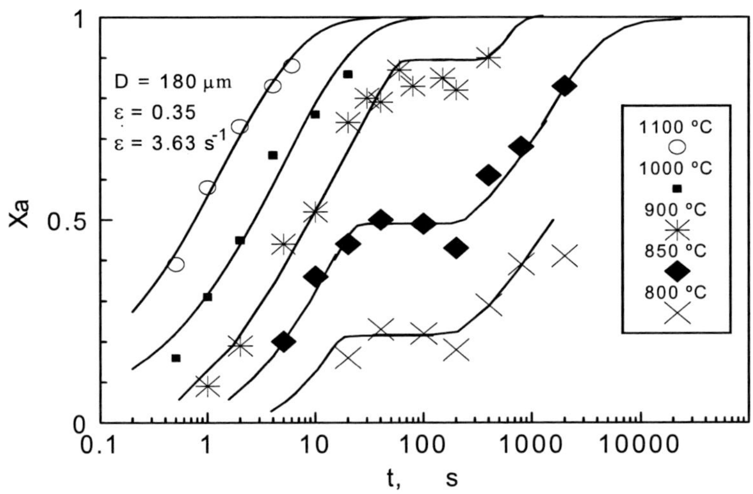

Figure 2. Recrystallized fraction (X_a) plotted against time (t) V-microalloyed steel (C=0.21; Si=0.20; Mn=1.10; V=0.062; N=0.0134, wt%).

Figure 3 shows the microstructure corresponding to the reheating temperature of 1200°Cx10min, strain of 0 35, testing temperature of 900°C and recrystallization time of 70 s, corresponding to the start of the plateau in Figure 2. With the assistance of the optical microscope an estimate was made of the recrystallized fraction in each case, by means of the observation of several fields. The values found (X_{am}) have been noted at the foot of each micrograph together with the recrystallized fraction (X_a) deduced from Figure 2. The good correspondence between the fractions determined by means of the 'back extrapolation' method and those estimated from observation of the microstructure again shows the efficiency of the method. In the same way, Figure 4 shows the microstructures corresponding to the strain of 0 35, temperature of 850°C and recrystallization times of 50, which correspond approximately to a time after the starting of the plateau [61].

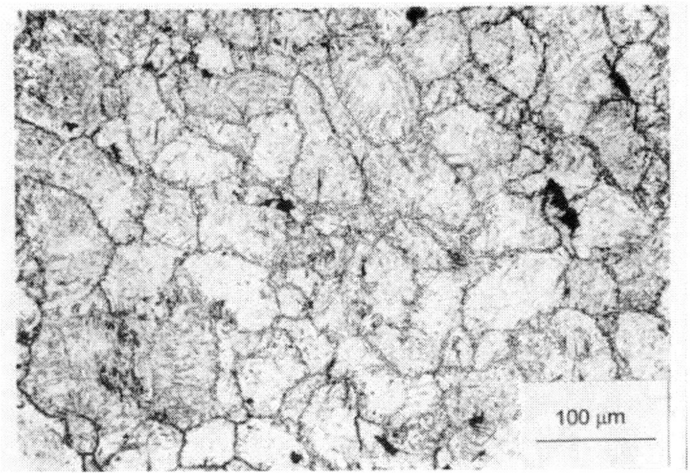

Figure 3. Microstructure of recrystallized fraction in austenite of steel used. Reheating temperature=1200°Cx10min; strain temperature=900 °C; $\varepsilon = 0.35$; $\dot{\varepsilon} = 3.63\ s^{-1}$; holding time before quenching = 70 s; $X_{am} = 0.85$; $X_a = 0.88$.

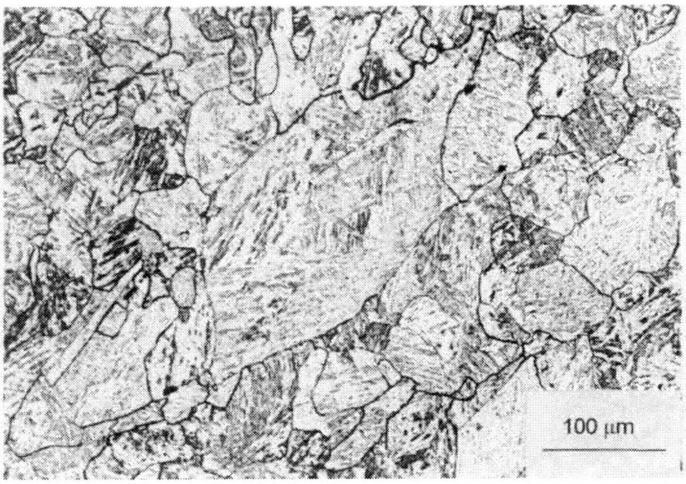

Figure 4. Microstructure of recrystallized fraction in austenite of steel used Reheating temperature=1200°Cx10min; strain temperature=850 °C, $\varepsilon = 0.35$; $\dot{\varepsilon} = 3.63\ s^{-1}$; holding time before quenching = 50 s; $X_{am} = 0\ 50$; $X_a = 0.52$.

The Model for Static Recrystallization Kinetics Proposed by Medina et al Is Divided in Two Submodels, when all the Elements Are in Solution [50,52] and when Induced Precipitation has Taken Place [51,53], Respectively

(i) For Microalloying Elements (V, Nb) in Solution

$$A = 3.754 \cdot 10^{-4} \exp(-7.869 \cdot 10^{-5} Q) \tag{19}$$

$$p = \text{-}4\ 3\ D^{-0\ 169} \tag{20}$$

$$q = -\ 0\ 535 \tag{21}$$

$$s=1$$

$$Q(J.mol^{-1}) = 1486368 - 719813[C] + 56537.6[Si] + 21180[Mn] + \\ + 1212433[Mo] + 64469.6[V] + 1097319[Nb]^{0.15} \tag{22}$$

All the steels, even V-microalloyed steels [52]:

$$n = 2.93 \exp\left(-\frac{12500}{RT}\right) \tag{23}$$

For Nb-microalloyed steels:

$$n = 28.33 \exp\left(-\frac{36000}{RT}\right) \tag{24}$$

(ii) The Strain Induced Precipitation Has Taken Place [53]

The activation energy (Q') for the static recrystallization of austenite with the presence of precipitates may be expressed as the sum of two terms:

$$Q' = Q + Q^* \tag{25}$$

where Q represents the activation energy in the absence of precipitates and Q^* represents the increase due to the presence of the precipitates.

The expression for the parameter $t_{0\ 5}$ for temperatures where the precipitation has taken place, or in other words applicable to austenite with the presence of strain-induced precipitates. This expression would take the following form:

$$t_{0.5} = A' \varepsilon^{(-4.28 D^{-0.17})} \dot{\varepsilon}^{-0.53} D \exp(\frac{Q'}{RT}) \tag{26}$$

To calculate Q, equation (22) is applied, bearing in mind that now neither vanadium nor niobium are in solution, but as precipitates, in accordance with the stoichiometry of vanadium nitrides (VN) and of niobium carbonitrides ($NbC_{0.7}N_{0.2}$), respectively. In the case of vanadium steels, only when the V content is high, above the stoichiometry ratio of VN, the vanadium content in solution does the value of Q rise slightly, as can easily be verified.

In accordance with equation (18), Figure 5 displays the parameter $\ln t_{0.5}$ against the inverse of the absolute temperature for a V-microalloyed steel, at two different austenite grain sizes corresponding to reheating temperatures of 1200°C and 1100°C, respectively. The slope of each line multiplied by the universal gas constant (R) gives the activation energy before (Q) and after precipitation (Q'). In other words, the value of the activation energy changes from one stage to another and this occurs at a constant temperature during the time that precipitation takes place. After precipitation the activation energy increases significantly, which is obviously translated into greater difficulty for the austenite to recrystallize.

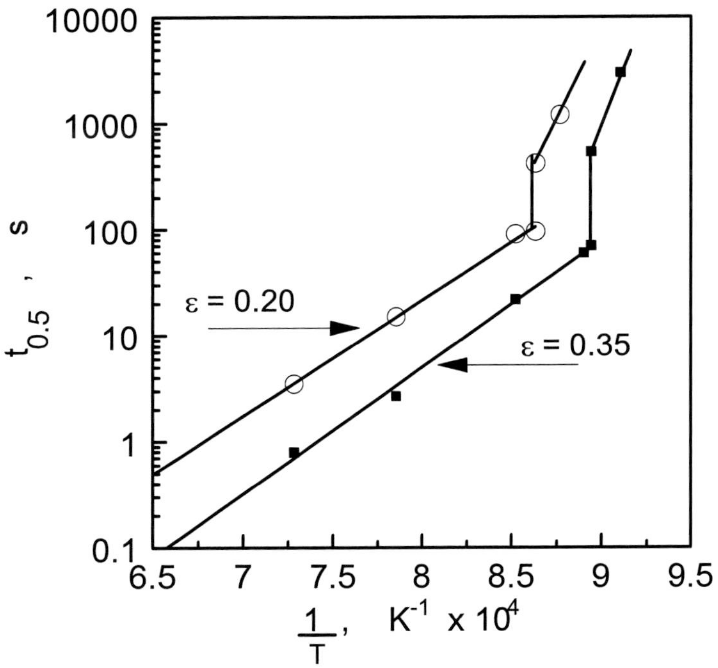

Figure 5. Plot of $t_{0.5}$ against the reciprocal of the absolute temperature for a V-microalloyed steel (C=0.094; Si=0.32; Mn=1 47; V=0.92; N=0.0065, wt%).

The expressions found for Q* were as follows:

V-steels: $$Q^*(J) = 1074 \cdot 10^3 [VN]^{0.223} \qquad (27)$$

Nb steels: $$Q^*(J) = 1577 \cdot 10^3 [NbC^{0.7}N^{0.2}]^{0.254} \qquad (28)$$

where Q' is given by equation (25), whose definitive expression would be:

$$Q'(J.mol^{-1}) = 148\,636.8 - 71981.3[C] + 56\,537.6[Si] + 21180[Mn] + \\ +121\,243.3[Mo] + 64\,469.6[V]_s + 109731.9[Nb]_s + 1074\cdot 10^3[VN]^{0.223} + \\ +1577\cdot 10^3[NbC^{0.7}N^{0.2}]^{0.254} \quad (29)$$

where $[V]_s$ and $[Nb]_s$ represent the percentages of vanadium and niobium that remain in solution, and which are calculated by subtracting from the total content the precipitated content, in accordance with the stoichiometric relation of vanadium nitrides and niobium carbonitrides, respectively.

The results showed that the coefficient A' does not depend on the deformation conditions ($\varepsilon, \dot{\varepsilon}, T$) or on the grain size, but on the activation energy Q′, finding the following general relation for all the steels studied:

$$A' = 1.2553\cdot 10^{-6}\exp\left(7.446\cdot 10^{-5}Q'\right) \quad (30)$$

where Q' is given in $Jmol^{-1}$

Finally, the parameters *p, q, s* and *n* will be the same in both submodels.

Recrystallization-Precipitation-Time-Temperature (RPTT) Diagrams

Recrystallized fraction curves can be used to plot recrystallization-precipitation-time-temperature (RPTT) diagrams. The points that define the start and end of the plateau were taken to plot the induced precipitation start (P_s) and finish (P_f) curves, respectively._The temperatures and times corresponding to different recrystallized fractions, such as 0.1, 0.3, 0.5, 0.9 and 0.95, were also deduced from recrystallized fraction curves. In this way we have drawn the RPTT diagrams –of which Figure 6 is an example– corresponding to a V-micraolloyed steel. Note that the recrystallized fraction does not vary between the precipitation start (P_s) and finish (P_f) curves and is represented by a horizontal line. Once the P_f curve is reached, the lines of each recrystallized fraction descend, meaning that as the temperature drops more time is necessary to obtain a certain recrystallized fraction after straining. With regard to the recrystallization-precipitation interaction, it is seen that at the nose of the P_s curve, where the incubation time of the precipitates (t_N) is minimal, the recrystallized volume fraction is approximately 50%. In other words, the optimum microstructure for precipitation to take place is that corresponding to the 50% recrystallized fraction, where saturation and diffusion reach a maximum. When the recrystallized volume fraction is less than 30%, nucleation of the precipitates is very difficult. In other words, for nucleation to be possible it is necessary for the recrystallized fraction to be equal to or greater than 30%. The most important magnitudes that can be deduced from an RPTT diagram, which are also indispensable for the perfect configuration of the diagram, are the aforementioned minimum incubation time (t_N), minimum precipitation end time (t'_N), curve nose temperature (T_N) and finally the precipitation duration time (t'_N-t_N). At the moment when precipitation starts, whatever the temperature (P_s curve), it is assumed that the precipitated fraction corresponds to a value of 5%.

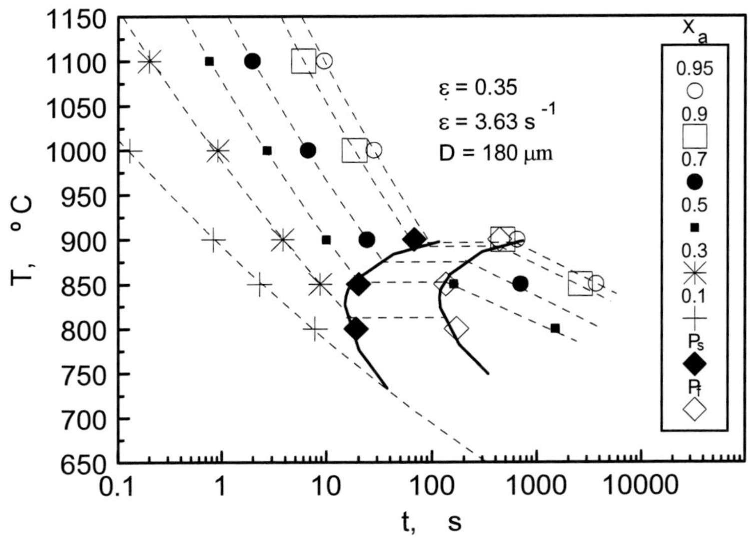

Figure 6. RPTT diagram for a V-microalloyed steel (C=0 21; Si=0.20; Mn=1.10; V=0.062; N=0.0134, wt%).

In the same way, when the P_f curve is reached, the precipitated volume is close to 95%. Once the P_f curve has been reached, recrystallization starts to progress again due to fact that the pinning forces exerted by the precipitates are lower than the driving forces for recrystallization. The RPTT diagrams, and especially the P_s and P_f curves, define a time interval, whatever the temperature, during which the precipitation state (size and precipitated volume) is changing. For times after P_f, the precipitated fraction does not vary but a coarsening of the precipitates occurs due to the effect of Ostwald ripening [57].

PRECIPITATES: SIZE AND DISTRIBUTION

The start of the plateau is identified with the start of precipitation and its duration is related with two phenomena that take place simultaneously; namely increases in the precipitated volume and in the size of precipitates. The latter effect is due to the heterogeneous nucleation of new precipitates. In this sense, several studies carried out with the assistance of transmission microscopy (TEM) have shown the speed with which the coarsening of precipitates occurs, while at the same time indicating how the size distribution changes from a lognormal distribution at the start of the plateau to a bimodal distribution at its end [55,56,58,59].

The coarsening of precipitates by coalescence occurs once the precipitation is complete, i.e. when the plateau is finished and the recrystallization progresses again. The coalescence can be explained by the modified Lifshitz-Slyozov-Wagner theory (MLSW) [60]. This theory predicts that while the basic $t^{1/3}$ kinetics of the MLSW theory is maintained, the coarsening rate increases with an increase in the volume fraction, even at very small precipitated volume fraction values. According to the MLSW theory, the theoretical distribution of particle sizes broadens rapidly at small volume fraction values.

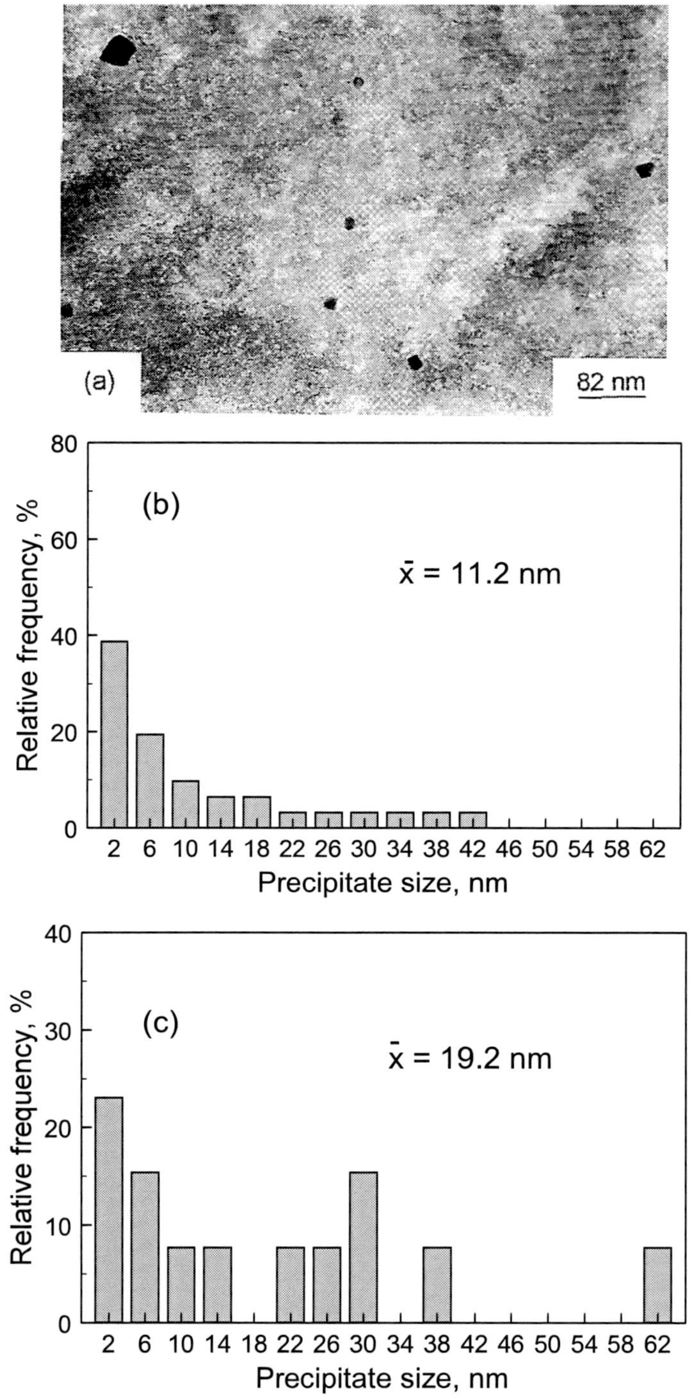

Figure 7. Precipitates and size distribution for a V-microalloyed steel (C=0.33; Si=0.22; Mn=1.24; V=0.076; N=0.0146, wt%) (a) TEM image showing VN particles; (b) Size distribution of precipitates for specimen tested at: reheating temp. = 1200°Cx10 min; test temp. = 900°C; ε=0.35; recrystallization time=50 s (start of plateau) (c) Size distribution of precipitates for specimen tested at: reheating temp = 1200°Cx10 min; test temp. = 900°C; ε=0.35; recrystallization time=250 s (end of plateau).

Figure 7a shows the TEM resolution of vanadium nitrides corresponding to V-microalloyed steel (C=0.33; Si=0.22; Mn=1.24; V=0.076; N=0.0146, wt%) obtained on specimens strained and quenched in the conditions indicated at the foot of the figures. Figure 7b shows a lognormal distribution of the precipitates with a weighted mean size of 11.2 *nm*, with the greatest frequency corresponding to a average size of 2 *nm*. However, Figure 7c shows a bimodal distribution with a weighted mean size of 19.2 *nm* and two relative maximum frequencies of 2 and 30 *nm*, respectively. Both micrographs serve to confirm that the end of the plateau is due to the coarsening of precipitates and consequently to a decrease in inhibition forces. Other studies carried out with Nb-microalloyed steels have shown similar results with regard to precipitate coarsening and also in relation to the two types of size distributions [57].

INTRAGRANULAR PRECIPITATION

In addition to the known classic sites for heterogeneous nucleation of ferrite, such as grain boundaries and dislocations, in recent years a third way is being studied which consists of the intragranular nucleation of ferrite on precipitates and inclusions. Some authors have studied the intragranular nucleation of acicular ferrite on MnS-CuS inclusions [62] and also on Ti_2O_3 particles [63-66]. Others have studied the intragranular nucleation of idiomorphic ferrite on complex particles formed by V(C,N) precipitates and inclusions [67-69].

The mechanisms that promote intragranular nucleation vary according to the type of particle or inclusion. On the one hand, the presence of MnS type inclusions causes a local Mn depletion that increases the A_{r3} value, favoring the nucleation of ferrite. On the other hand, VC or VN particles have a low interfacial energy in relation to ferrite but a relatively high interfacial energy in relation to austenite for the $(001)_{V(C,N)}$ boundary compared to MnS. These advantages of VC and VN over MnS in the interphase boundary energy balance presumably promote the intragranular ferrite transformation for complex precipitates [70].

However, the intragranular nucleation of ferrite has not yet been clearly demonstrated and its contribution to ferrite grain refinement is still to be quantified. The present work represents the continuation of previous work by the authors on induced precipitation and on the intragranular nucleation of ferrite in vanadium microalloyed steels, using our own methodology developed over recent years, which seeks to demonstrate and to quantify the contribution of intragranular nucleation of ferrite on V(CN) precipitates [71,72].

Figures 8 and 9 show RPTT diagrams for two strains of 0.20 and 0.35, respectively, for a low carbon V-microalloyed steel (C=0.103; Si=0.304; Mn=1.463; V=0.139; N=0.0100, wt%), and a series of arrows indicate the holding times at 950°C, after deformation, in tests performed with the aim of studying the influence of precipitation on intragranular nucleation, as will be seen in the following section. According to the RPTT diagrams, the deformation conditions were selected to coincide approximately with times before the nose of the P_s curve, close to P_s and P_f, and after P_f, respectively. In this way it was possible to evaluate the effect of precipitation kinetics on the intragranular nucleation of ferrite.

The aim of these tests is to verify whether two austenitic microstructures subjected to the same deformation at a certain temperature, with holding times after deformation situated between the P_s and P_f curves, equivalent to a plateau, and cooled at the same cooling rate, would yield the same or a different ferrite grain size. It should be noted that two or more holding times corresponding to the P_f-P_s interval would have the same microstructure, i.e. the same recrystallized fraction, which means the same dislocation density and the same average austenite grain size, and the only difference would be the precipitated volume.

Both microstructures will be differentiated by the precipitated volume, conventionally from 5% corresponding to P_s, which corresponds to the start of the plateau, and 95% corresponding to P_f.

The deformation conditions are shown in Table 1. After the holding time at the deformation temperature, the specimens were cooled at a rate $(dT/dt)_{800\text{-}500°C}$ of 3.5 K/s. Table 3 also reports the recrystallized fraction and the precipitated volume, calculated in accordance with the RPTT diagrams.

Metallographic preparation of the specimens allowed the measurement of the ferrite grain size, as is shown in the ferritic microstructures obtained in the conditions noted below (Figure 10 a-g). Ferrite grain sizes were graphically represented as a function of the austenite holding time after the deformation temperature (Figure 11). An increase in the holding time (Δt) after deformation reduces the ferrite grain size, but is insufficient to suggest that grain refinement has been due exclusively to VN precipitates acting as nucleation sites for ferrite during cooling.

Table 1. Testing conditions: Deformation temperature (T_d), strain (ε), recrystallization / precipitation time (Δt), recrystallized fraction (X_a), precipitated fraction (X_p) and ferrite grain size (D_α)

T_d, °C	(ε)	Δt, s	X_a, %	X_p, %	D_α, µm
950	0.20	10	20	0	21
		50	53	5	17
		700	53	90	13.5
950	0.35	10	48	0	20
		20	55	5	15.5
		100	55	75	13.5
		500	73	100	12.5

As has been noted, when austenite is deformed at a given temperature and held for a certain time at that temperature, the resulting microstructure can recrystallize either partially or completely. A hardening of austenite, originated by incomplete recrystallization, increases the density of dislocations and consequently the number of nucleation sites. Therefore, the contribution of intragranular nucleation on precipitates to ferrite grain refinement must be evaluated in conditions where both the austenite grain size and the dislocation density remain constant. In this sense, it is supposed that two austenitic microstructures would have the same dislocation density if the recrystallized fraction is the same.

The recrystallized fractions and precipitated fractions may be deduced from the RPTT diagrams (Figures 8 and 9), making the opportune interpolations. Thus, for a strain of 0.20 the values of X_a (Table 1) corresponding to holding times (Δt) of 50 and 700 s were 53%. The precipitated fractions for these times were 5%, coinciding with P_s, and 90%, slightly before P_f (Figure 8), and thus close to the start and the end of precipitation. Between these two conditions, both the average austenite grain size and the dislocation density are approximately the same and only the precipitated fraction varies. The latter has been calculated bearing in mind that the precipitation kinetics between P_s and P_f obey Avrami's law [73]. Nevertheless, a linear approximation between P_s and P_f would yield very similar results.

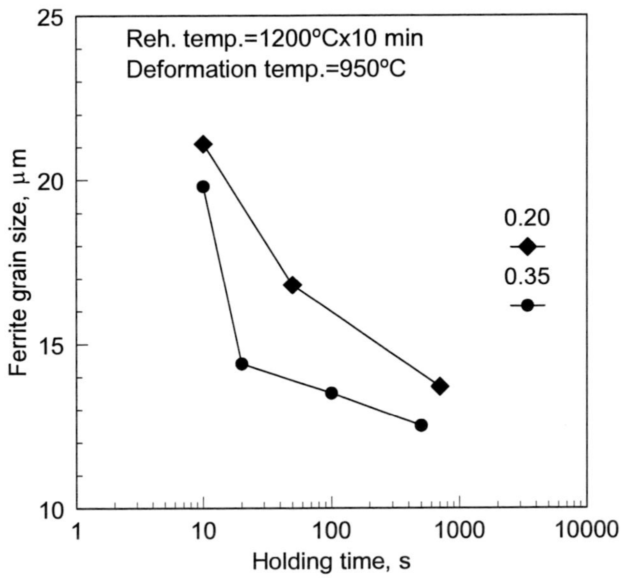

Figure 8. Ferrite grain size against holding time alter deformation.

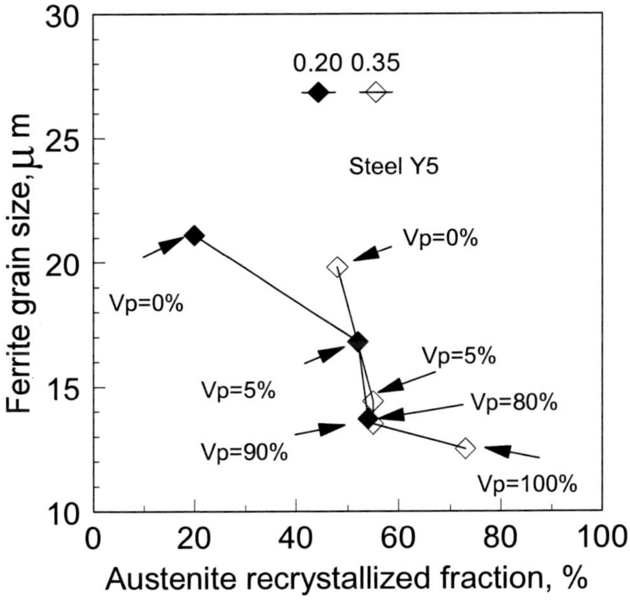

Figure 9. Ferrite grain size against austenite recrystallized fraction.

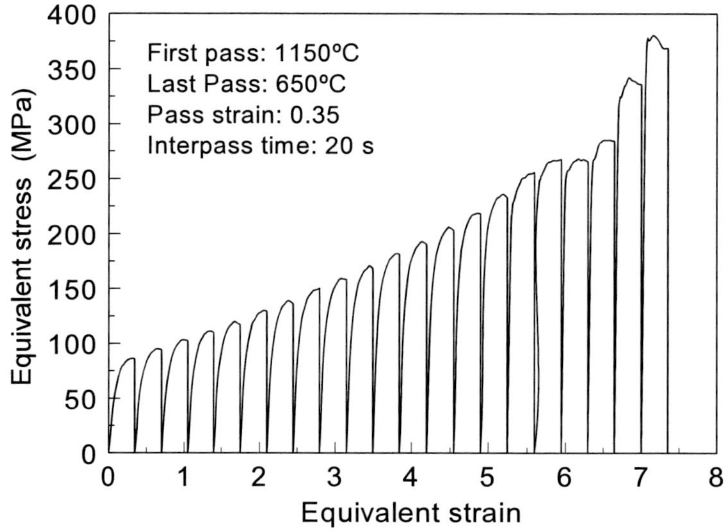

Figure 10. Strain-stress curves in hot rolling simulation.

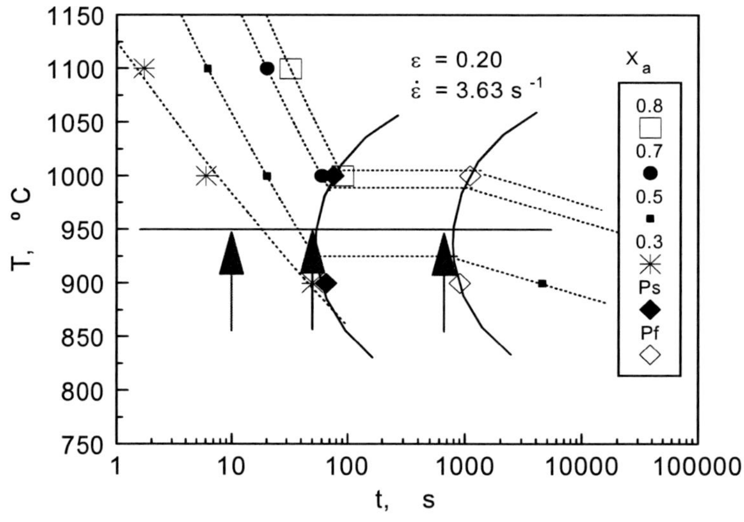

Figure 11. RPTT diagram. Strain=0.20.

These results are shown in Figure 12, where the ferrite grain size has now been represented as a function of the recrystallized fraction, related with the dislocation density, noting on each point the corresponding precipitated fraction. The contribution of ferrite intragranular nucleation would be measured on the approximately vertical segment of each representation corresponding to the strains of 0.20 and 0.35, respectively, if necessary making extrapolations to the 0% and 100% precipitated fraction, respectively.

The reduction in the ferrite grain size (D_α) obtained between the two aforementioned conditions (holding times of 50 s and 700 s) was 3.5 μm, which extrapolated to a 95% precipitated fraction would mean approximately 4 μm, which compared to the grain size corresponding to a 5% precipitated fraction (D_α = 17 μm) implies a 24% reduction. In the same way, for a strain of 0.35 and holding times (Δt) of 50 and 700 s, the X_a values were 55%. In these conditions of the same average austenite grain size and the same dislocation density, the precipitated volume varies between 5% and 80%. Between these conditions the reduction obtained for D_α was 2 μm, which extrapolated to a 95% precipitated fraction would mean 2.5 μm, representing a reduction of approximately 16% from the size corresponding to a 5% precipitated fraction (D_α = 15.5 μm).

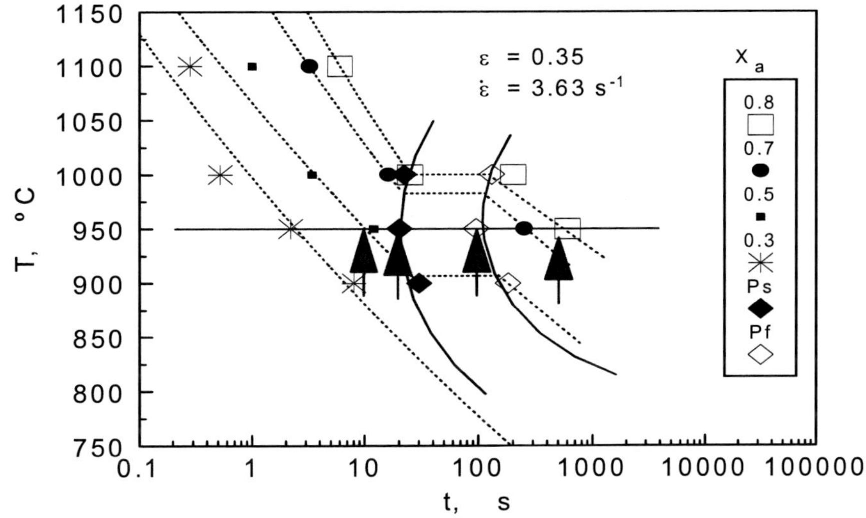

Figure 12. RPTT diagram. Strain=0.35.

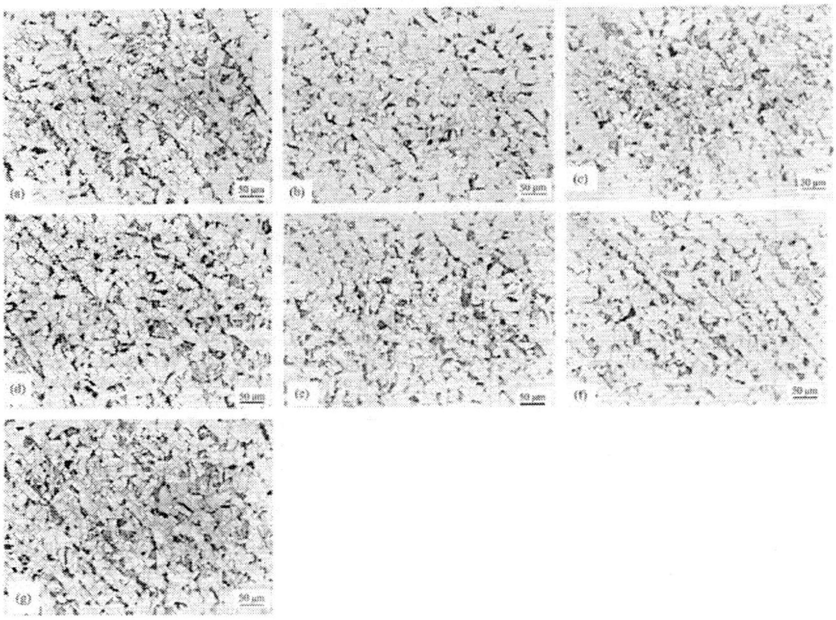

Figure 13. Ferrite-pearlite microstructures T_d=940°C; Cool Rate=3.5K/s; ε=0.20: (a) Δt=10 s, (b) Δt=50 s, (c) Δt=700 s; ε=0.35: (d) Δt=10 s, (e) Δt=20 s, (f) Δt=100 s; (g) Δt=500 s.

Comparing the above results it is seen that the recrystallized fraction between P_s and P_f is approximately 53% and 55% for strains of 0.20 and 0.35, respectively. This means that both the recrystallized and the non-recrystallized fractions are similar, and in particular that the austenite hardening reached in the latter case on the non-recrystallized fraction is obviously greater. On the other hand, the recrystallized grain size decreases as the strain increases. Thus it may be concluded that in the case of the strain of 0.35, the dislocation density was greater and the average austenite grain size was smaller than with the 0.20 strain. Both aspects favor the nucleation of ferrite, both at austenite grain boundaries and on subgrains formed by the stacking or cluster of dislocations [74]. Therefore, it was expected that the contribution of intragranular nucleation to ferrite grain refinement would be less in the case of the 0.35 strain than in the case of the 0.20 strain, as has indeed been the case.

On the other hand, it is not possible to measure the percentage contribution of each of the nucleation sites separately, since as the recrystallized fraction increases so the average grain size decreases, favoring nucleation at the grain boundaries, and at the same time the dislocation density decreases, reducing the capacity of dislocations to serve as nucleation sites for ferrite. This is observed very well when the holding time after the 0.20 strain increases from 10 s to 50 s (Figure 11), causing an increase in the recrystallized fraction from 20% to 53% (Figure 12) and an undetermined reduction in the dislocation density.

Finally, in other work reported by the authors [75] it was shown that in V-microalloyed steels with a greater precipitated volume fraction, which is obviously achieved with higher V and N contents, the percentage reduction in the ferrite grain size due to intragranular nucleation on VN precipitates rises notably, and in a steel with C=0.1; V=0.151 and N=0.0227 (%) a 45% reduction was achieved. In that case the total precipitated fraction at the deformation temperature was 1.1×10^{-3}, more than one order of magnitude higher than in the present case [76].

Refinement of Ferrite Grain in Thermomechanical Treatment

The no-recrystallization temperature (T_{nr}) represents the start of inhibition of the static recrystallization of austenite during hot rolling. The most common method for determining T_{nr} consists of simulating successive rolling passes and then graphically representing the Mean Flow Stress (MFS) versus the inverse of the absolute temperature for each of the simulated passes.

The inhibition of recrystallization indicated by T_{nr} appears as a rise in the slope of the MFS curve and in the studied steel is mainly caused by the strain-induced precipitation of niobium carbonitrides [77-80].

This method also offers the possibility of obtaining the A_{r3} phase transformation temperatures when austenite is being deformed in conditions similar to rolling and the stress accumulated ($\Delta\sigma$) in the austenite between T_{nr} and A_{r3} [81]. The accumulated stress in austenite is a direct and precise measurement of the state of austenite as a function of the temperature and can be directly measured without the need to measure the size and lengthening of the austenite grain [82].

In the hot working of microalloyed steels, all three of the aforementioned nucleation sites can coexist and act simultaneously on ferrite grain refinement. In this chapter the influence of different rolling conditions on the final ferrite grain size, varying the interpass time and the strain in each pass, has been studied [83].

The torsion test gives the applied torque versus the number of turns made on the specimen, which are transformed respectively into the equivalent stress and strain. Simulations consisted of 21 passes, with first pass temperature of 1150°C, a temperature drop of 25°C between each pass, and the final pass temperature of 650°C, were carried out. Several interpass times (Δt) were tested. The effect of the strain (ε) was also tested by applying values of 0.20 and 0.35 in two simulation series, respectively The strain rate ($\dot{\varepsilon}$) of 3.63 s^{-1} was applied in all the simulation series.

Figure 13 shows the simulation of 21 rolling passes for a V-microalloyed steel (C=0.165; Si=0.532; Mn=1.560; V=0.170; N=0.0160, wt%) with an interpass time of 20 s. The stress rises as the temperature decreases, and there is a change in the slope with a growth in the stress, which means a greater tendency toward strengthening. Later, the stress drops and then grows again during the final passes.

The meaning of these zones is better explained by observing Figure 14, which shows the graphic representation of Mean Flow Stress (MFS) versus the inverse of the absolute temperature for an interpass time of 20 s and strain of 0.35. MFS is determined in each step by dividing the area below the stress-strain curve by the strain applied.

It is possible to see three different zones. In the first zone (I), which corresponds to deformations at the highest temperatures, MFS grows as the temperature decreases. Austenite recrystallizes completely between passes and there is no accumulated stress. The increase in stress is only due to the decrease in temperature. In the second zone of the curve (II) there is a change in the slope, which indicates a greater tendency towards hardening. Here the stress accumulates in the austenite, whose recrystallization between passes is partially inhibited. The third phase (III), characterized by a drop in MFS as the temperature decreases, corresponds to the austenite to ferrite and pearlite transformation.

The intersection of the straight regression lines of phases I and II defines the value of T_{nr} and the intersection of the regression lines of phases II and III determines the value of A_{r3}. The method applied considers that if the austenite were to recrystallize completely between passes until reaching A_{r3}, the slope of the zone II straight regression line would be the same as that of the zone I straight line, and in this case the accumulated stress ($\Delta\sigma$) would be nil.

The value of $\Delta\sigma$ can be measured from the graph of MFS versus the inverse of the temperature and will be given by the length of the vertical segment limited by the phase I and phase II regression lines, as illustrated in Figure 14. When the temperature is A_{r3}, $\Delta\sigma$ reaches its maximum value in austenite. In this way it has been possible to measure accumulated stress for all the thermomechanical simulations in a quick and precise way, since this magnitude represents better than any other the degree of strengthening of the austenite in the temperature interval T_{nr}–A_{r3}.

Values of T_{nr}, $\Delta\sigma$ and A_{r3} were measured on the graphs of MFS versus 1/T for several tested deformation conditions. Figures 15 and 16 illustrate the influence of Δt and the strain on T_{nr} and $\Delta\sigma$, respectively T_{nr} decreases considerably with Δt and is always lower for the strain of 0.35. An increase in Δt obviously means improving the possibility of the recrystallized fraction increasing between passes with temperatures below T_{nr}.

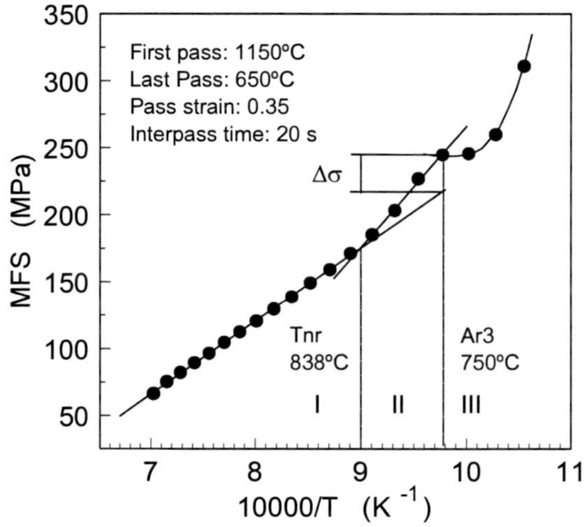

Figure 14. Dependence of mean flow stress (MFS) on inverse absolute temperature, illustrating as the accumulated stress (Δσ) is measured. Three phases (I-III) can been seen.

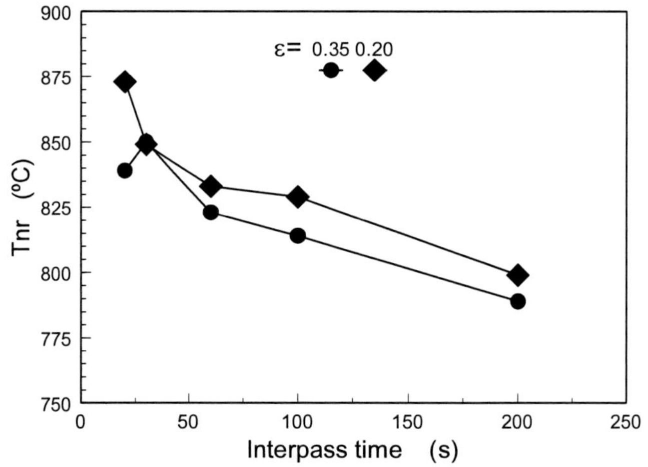

Figure 15. No-recrystallization temperature (T_{nr}) against interpass time.

In the same way, Δσ decreases with Δt and is also lower when the applied strain was 0.35. These results indicate that in order to achieve good strengthening of the austenite during hot rolling, close to pancaking, it is necessary to apply low strains and to reduce the interpass time, as would be the case in the finishing strip mill. In plate hot rolling it is more difficult to achieve pancaking, but a reduction in the interpass times and low strains in the final passes would help to achieve this objective.

Following the geometric procedure noted above for determining Δσ, the accumulated stress can easily be measured at any temperature above A_{r3} and below T_{nr}, thus making it possible to know the gradual grade of strengthening as a function of the temperature. For this purpose it is sufficient to subtract from the zone I regression plot the zone II regression plot for temperatures between T_{nr} and A_{r3}. The results calculated in this way are shown in Figures 17 and 18 for strains of 0.20 and 0.35, respectively.

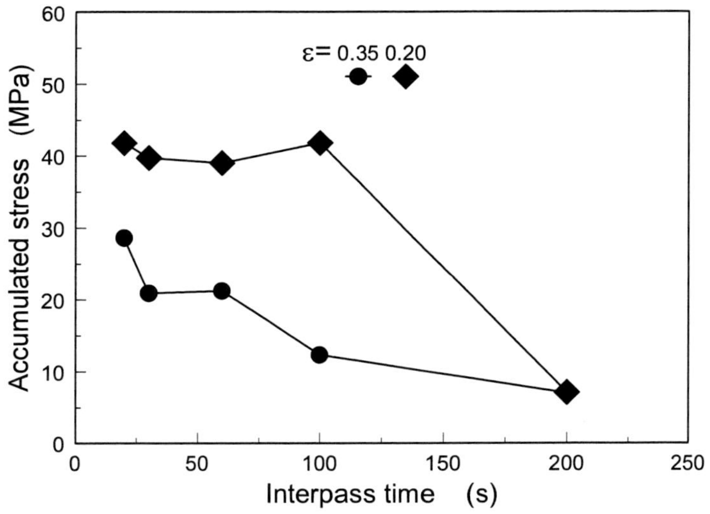

Figure 16. Accumulated stress at A_{r3} against interpass time.

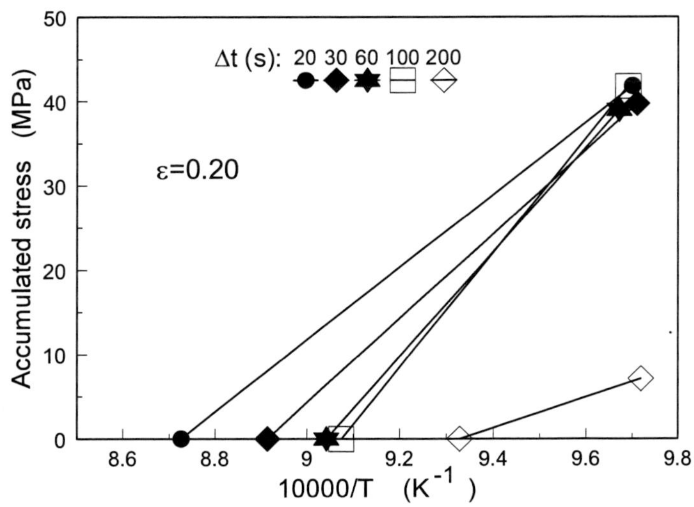

Figure 17. Accumulated stress against reciprocal temperature in hot rolling simulation. Pass strain =0.20.

Given that the hardening of austenite takes place between T_{nr} and A_{r3}, the difference between these two temperatures (ΔT) marks the interval where hardening is possible. The influence of strain was not important and only a regression was done. The logarithmic regression, whose curve has been plotted, yielded a higher correlation index compared to other possible regressions, such as potential or exponential, whose expression is as follows:

$$\Delta T\ (^{\circ}C)=184.5-26.8\ \ln(\Delta t) \tag{31}$$

Comparison of these results with Nb-microalloyed steels indicates that this interval is much smaller for V-microalloyed steels, even with much greater V than Nb contents.

However, the interval for V microalloyed steels is greater than that for Ti microalloyed steels, where the T_{nr} temperature is often inexistent when the rolling conditions imply a relatively high pass strain and even when interpass times are short [84].

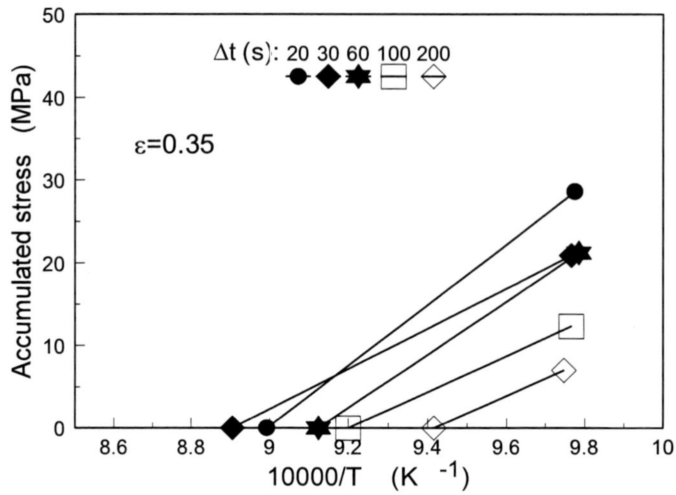

Figure 18. Accumulated stress against reciprocal temperature in hot rolling simulation. Pass strain =0.35.

Microalloyed steels are normally rolled at temperatures above A_{r3} and the final pass, e. g. in the hot strip mill, is performed at a temperature close to $A_{r3.}$ In view of this fact, new simulations were carried out in the same conditions as the previous simulations but performing the last pass at 800°C and cooling the specimen in an argon stream at a rate of approximately (3.5 K/s) $_{(800\text{-}500)°C}$ to room temperature. Figures 19 a-b show microstructures obtained in the conditions noted in the respective captions with regard to pass strain and interpass time. The ferrite grain sizes measured (Dα) for all the conditions are shown in Figure 20 versus the accumulated stress (Δσ) in austenite at 800°C. On the other hand, the accumulated stress at 800°C interprets the stress state of austenite at that temperature and is closely related with the dislocation density [85].

a) 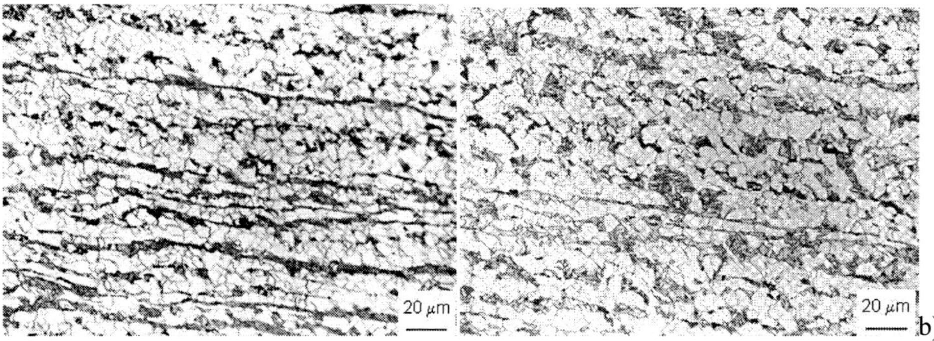 b)

Figure 19. Microstructures obtained by hot rolling simulation and cooling from 800°C. Pass strain=0.20; interpass time: (a) 20 s; (b) 100 s.

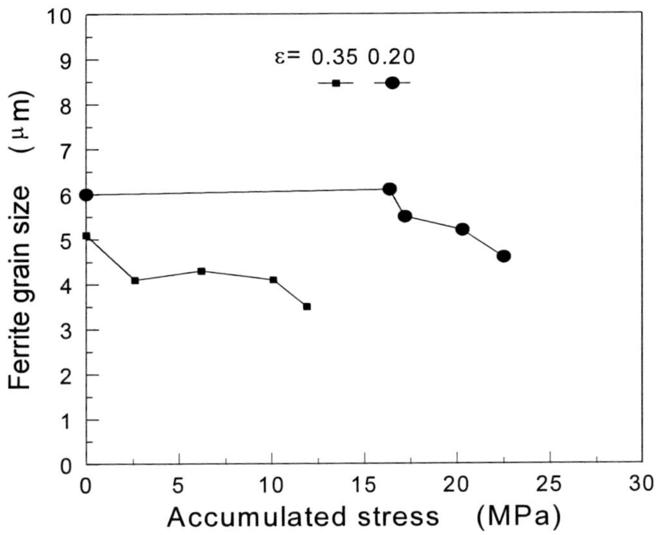

Figure 20. Ferrite grain size against accumulated stress.

Although larger values of $\Delta\sigma$ were reached when the strain was 0.20, D_α was lower for the strain of 0.35. The increase in strain originates a greater recrystallized fraction between passes and thus a lower $\Delta\sigma$ value. At the same time, the average austenite grain size will be somewhat smaller when the strain is greater, because a higher recrystallized fraction percentage reduces the average austenite grain size [24]. In the case of the strain of 0.20, Dα starts to drop slightly when $\Delta\sigma$ exceeds a value of 17 MPa. For the strain of 0.35, Dα shows a slight tendency to decrease with the accumulated stress, although not so clearly as in the previous case.

On the other hand, it is very well known that in order for the nucleation of ferrite to take place to an important extent on arrangements of dislocations (subgrains) it is necessary to reach an accumulated stress of more than 40 MPa. Consequently, the nucleation of ferrite has happened preponderantly on grain boundaries, as can be deduced by comparison of the graphs corresponding to the two strain values, and possibly also on VN precipitates. Nevertheless, it is difficult to quantitatively distinguish these two contributions from the tests performed.

The influence of the austenite grain size prior to the ferrite transformation is known, but in the present case of a V-microalloyed steel with relatively high V and N contents it is useful to know, albeit approximately, the quantitative relation between the two magnitudes. For this purpose simulations were performed at a pass strain of 0.20, an interpass time of 30 s, and final rolling temperatures of 1100°C, 1000°C, 900°C and 800°C, respectively. In the first series of simulations each specimen was quenched in water after being deformed at the final rolling temperature and held for a time of 30 s at this temperature. In this way the accumulated stress reached in the austenite after holding may be deduced from Figure 16. In the second series the specimens were cooled by means of an argon stream at a cooling rate of approximately 3.5 K/s. The measured grain sizes are shown in Figure 21. The final temperature of 800°C is lower than T_{nr}, which corresponds to a $\Delta\sigma$ value of approximately 19 MPa. However, the accumulated stress corresponding to the final temperatures of 900, 1000 and 1100°C is zero, because these temperatures are higher than T_{nr}.

Therefore, in Figure 21 the point referring to the final temperature of 800°C, represented by an asterisk, is outside the logarithmic tendency of the other points which are related to completely recrystallized austenites, i. e. austenites without deformation. It should be noted that the average austenite grain size was approximately the same at 900°C as at 800°C. If it is accepted that strain induced precipitation in microalloyed steels starts at a temperature close to T_{nr}, it is highly likely that the ferrite grain refinement obtained by cooling from 800°C compared to that obtained in cooling from 900°C is not only due to the accumulated stress but also to the possible intragranular nucleation of ferrite on VN precipitates.

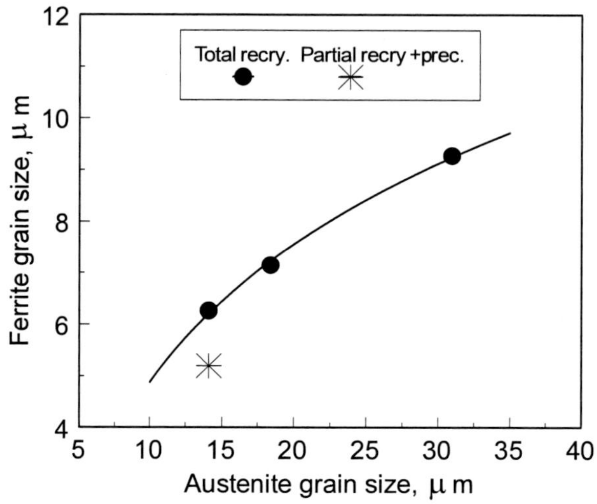

Figure 21. Ferrite grain size against to austenite grain size.

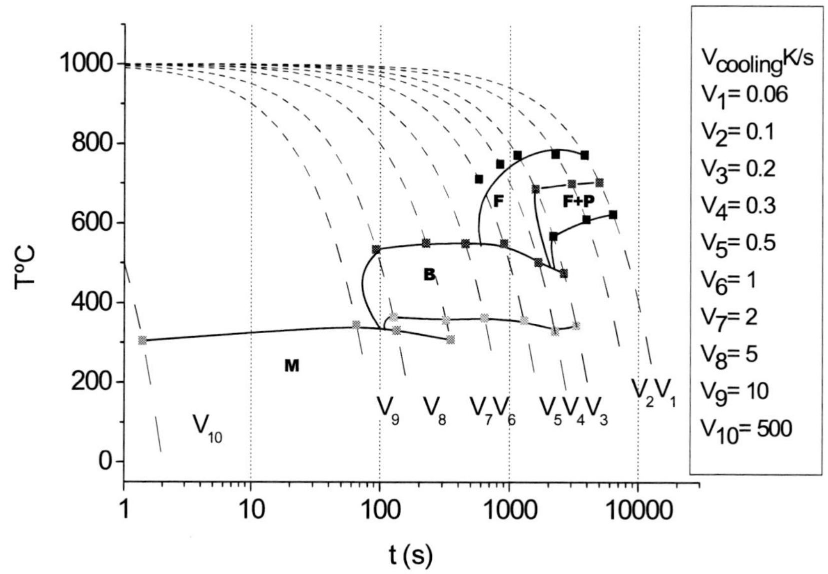

Figure 22. Cooling continuous transformation (CCT) diagram for medium carbon V-microalloyed steel.

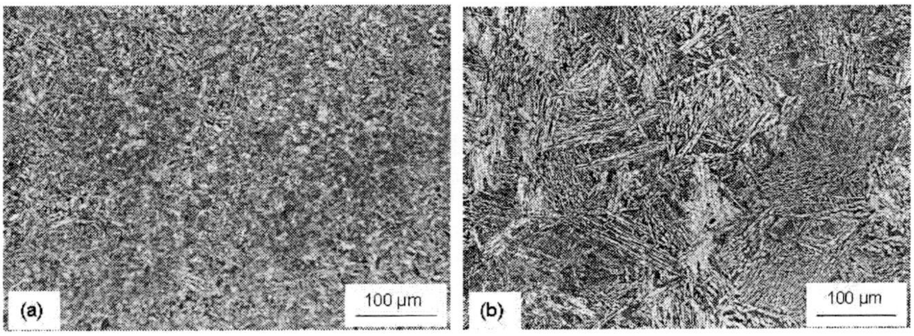

Figure 23. Bainite microstructures for medium carbon V-microalloyed steel: (a) cooling from 950°C; (b) cooling from 1150°C.

Mechanical Properties

The Ashby-Orowan relationship has been used on steels with ferrite-pearlite microstructure and its application is very well-known [3]. However, its application on bainitic steels is not wide spread [3]. Recent developments have been achieved on Nb and V-microalloyed bainitic steels, with low and medium carbon contents, respectively. In this regard, the results of the latest works about V-microalloyed steels are very interesting [86,87].

The Orowan and Ashby-Orowan expressions forecast the strengthening effect of a dispersion of fine particles. They represent an overestimate of the stress required to cause dislocation bypassing of the particles, so its prediction on ferrite and bainite could be different. The refinement in particle size produces significant increases in yield strength in accordance with both expressions [3]:

Orowan:

$$\Delta\sigma = \frac{83.2 \cdot f^{1/2}}{X} \tag{32}$$

Ashby-Orowan:

$$\Delta\sigma = \frac{10.8 f^{1/2}}{X} \ln(\frac{X}{6.125 \cdot 10^{-4}}) \tag{33}$$

where, X is the true three dimensional average diameter of particles in µm, and f is precipitated volume fraction. The relationship between X and the diameter x of particles of mean area in the plane of intersection is given by:

$$X = (3/2)^{1/2} x \tag{34}$$

Equation (32) represents an upper bound to the strengthening effect of a dispersion of fine particles, as it implies that the dislocation is parallel with one edge of the assumed square array of particles in the slip plane.

Aiming to know the predictive ability of (33) and (34) equations, the precipitate volume fractions (*f*) were calculated for a medium carbon V-microalloyed steel (C=0.38; Si=0.31; Mn=0.93; Cr=1.06; Mo=0.21; V=0.11; N=0.0170, wt%) taking into account the Turkdogan solubility products and according to the Manohar et al procedure, respectively [1,88]. Since in this steel the ratio V/N exceeds the vanadium nitride (VN) stoichiometric relationship, it was also calculated the precipitated volume fraction of carbides ($VC_{0\ 75}$), being the whole fraction of precipitated volume the sum of nitrides and carbides. The precipitated volume is supposed to be the same that corresponds to equilibrium, before reaching the start temperature of the bainite transformation, which was approximately 550°C as it is deduced from the continuous cooling transformation (CCT) diagram (Figure 22).

Table 2 shows the obtained results for $\Delta\sigma$ predicted by equations (1) and (2) as well as the experimental ones at two austenitization temperatures. The bainite microstructures are shown in Figures 23 a-b. Figure 24 shows SEM-FEG micrograph where nanometric V-precipitates (VN) are seen together with bright laths of cementite. The precipitates were found inside the ferrite laths, which have a darker appearance, and this gave rise to the strengthening. The mean sizes for precipitates were of 37.6 nm and 37.8 nm at the austenitization temperatures of 950°C and 1150°C, respectively.

Table 2. Predicted and experimental strengthening ($\Delta\sigma$) for steels CR2 y CR6

Aust Temp, °C	f	x, (nm)	X, (μm)	Δσ, Equat 1, (MPa)	Δσ, Equat 2, (MPa)	Δσ, experimental (MPa)
950	$1.68{\cdot}10^{-3}$	37.6	$4.6{\cdot}10^{-2}$	74	42	79
1150	$1.68{\cdot}10^{-3}$	37.8	$4.6{\cdot}10^{-2}$	73	41	64

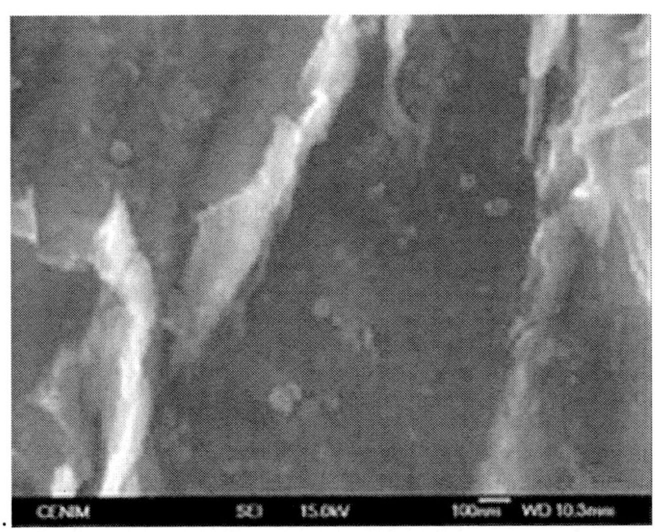

Figure 24. SEM image showing VN-nanometric particles Steel CR6 950°C.

The Orowan equation (1) indicates a significantly larger increase in yield strength than is forecast by the Ashby-Orowan equation (2), and this is due to the use of a minimum spacing of a regular array of particles in the Orowan equation as compared with effective spacing of the random array of particles in the Ashby-Orowan equation.

The experimental yield strength increase ($\Delta\sigma$) was obtained calculating the difference between the value which corresponds to reference steel, without V, but with the same content in the others alloy elements. Austenite grain size of this steel at 950°C and 1150°C was similar to the grain size of reference steel at the same respective temperature. Therefore, the differences found in yield strength values will be due to precipitation strengthening. The experimental values are near those predicted by the Orowan equation (1). In any case, the precipitation strengthening owing to the V-particles was approximately 70 Mpa.

On the other hand, 38MnV7 steel (C=0.38; Si=0.25; Mn=1.53; V=0.11; N=0.0217, wt%) is used to manufacture automotive parts and reaches a strength of approximately 1000 MPa. However, toughness values recorded in Charpy impact tests vary strongly depending on whether the bainitic microstructure is obtained after continuous cooling (upper bainite) or by means of isothermal treatments (lower bainite) [87]. The strengthening mechanisms that operate in bainite are well known, and a small bainite packet size means a small lath width, low dislocation density and a low number of carbide particles (Fe_3C) [89-92]. These properties are more easily achieved for lower bainite than for upper bainite.

The thermal treatments applied to the 38MnV7 steel specimens were as follows:

Austenitization temperature: 820°C / 30 min
Bainitic transformation temperature: 360°C / 90 min
Tempering temperature: 600°C / 60 min

The microstructure obtained after isothermal treatment showed a lower bainite (Figure 25), much finer than that obtained in anisothermal treatment which yields a coarser higher bainite with a larger bainitic packet [88]. The toughness of this steel may be considered excellent even in the case of isothermal treatment without tempering (Figure 26).

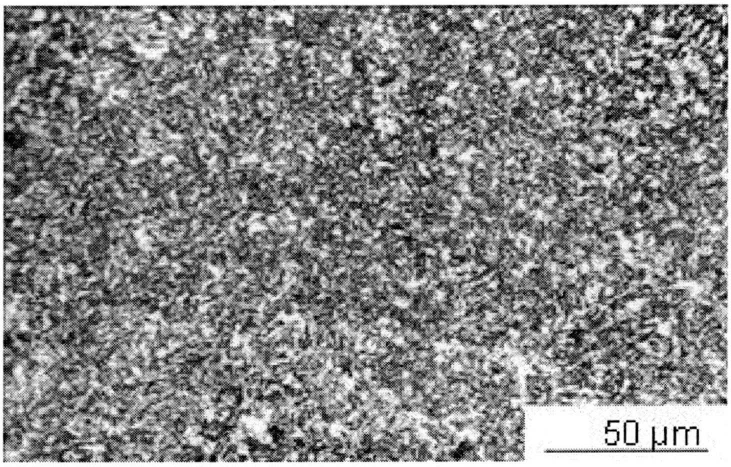

Figure 25. Bainite microstructure. Heat treatment: Aust. Temp. = 820°C; Transf. Temp. = 360°C Steel 38MnV7.

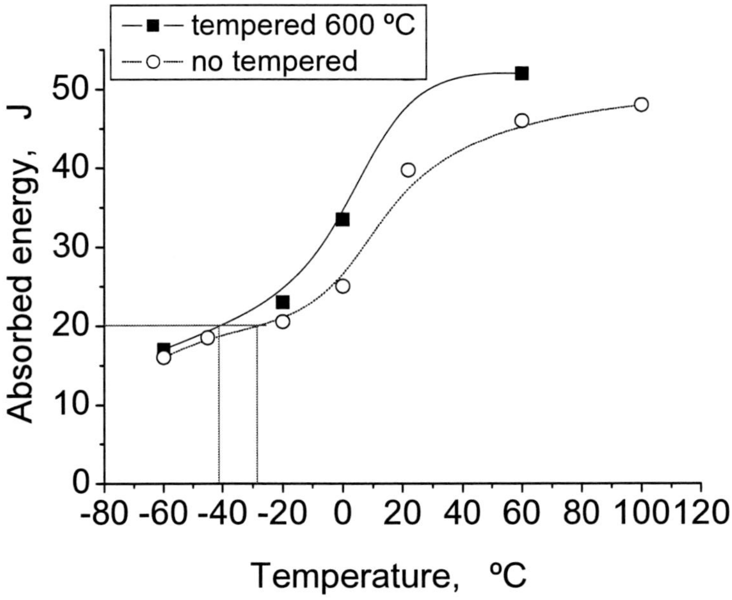

Figure 26. Charpy impact absorbed energy. Steel 38MnV7.

The steel presented a impact transition temperature (ITT) below 0°C for an absorbed energy of 20 J. Specifically, the steel presents an ITT of -29°C when not tempered and -41°C when tempered. These values indicate that steel 38MnV7 can be used in the most adverse service conditions even in a not-tempered state. The relatively high content of N to reach a V/N ratio near 5, and therefore a high precipitated volume fraction of V-particles (nitrides, carbides) contributed to enhance the intragranular nucleation of bainite and thus also the toughness.

CONCLUSION

The model for dynamic recrystallization makes it possible to calculate the recrystallized fraction of any low alloy or microalloyed steel as a function of the strain, temperature, strain rate and austenite grain size. All the alloying elements, except carbon, tend to delay the start of dynamic recrystallization. The flow stress model also makes it possible to understand the evolution of the microstructure during rolling when the pass strain is relatively high.

Before the start of the plateau and after its finish, static recrystallization kinetics obeys Avrami's law. Activation energy is the parameter that basically reflects the influence of precipitation on the delay experienced by static recrystallization kinetics. Activation energy in the precipitation state reaches values that are between two and three times those corresponding to the previous state, when all the elements are in solution. The model provides good prediction for V-microalloyed steels. The model constructed in isothermal conditions, may be applied in hot rolling by means of its conversion to continuous cooling conditions, e. g. using the compensated times method, and applying it to temperatures below that at which precipitation starts (T_{nr}).

The contribution of ferrite intragranular nucleation on V(C,N) particles is quantitatively smaller than the contribution of nucleation at grain boundaries and on dislocation stacking (subgrains). Nevertheless, in the present study it has been seen that the ferrite grain size was notably reduced by the intragranular nucleation of ferrite on precipitates. Intragranular nucleation led to a reduction of approximately 25% in the ferrite grain size with a strain of 0.20 and 16% with a strain of 0.35. In accordance with the above, the contribution of intragranular nucleation decreases as the applied strain increases. An increase in the strain favors nucleation of ferrite at austenite recrystallized grain boundaries, as the austenite grain is finer, and on subgrains, in preference to intragranular nucleation on VCN-precipitates.

The accumulated stresses between T_{nr} and A_{r3} were relatively small as a consequence of the low T_{nr} values obtained, implying an insufficient temperature interval to reach high $\Delta\sigma$ values that favor ferrite grain refinement by intragranular nucleation on dislocations. The accumulated stress is a direct and precise measurement of the state of austenite as a function of the temperature, and is greater the lower the strain applied in each pass. In order to obtain strongly deformed austenite microstructures before the $\gamma \rightarrow \alpha$ transformation it is advisable to reduce the interpass time and the magnitude of the strain applied in the final passes.

The V-precipitates (nitrides, carbides) increase the yield strength of bainitic steels and experimental values of strengthening were close to those predicted by Orowan equation. The V-precipitates also contribute to promote the intragranular nucleation of bainite, thus refining the bainitic packet and improving the toughness.

REFERENCES

[1] Turkdogan, E T *Iron Steelmaker* 1989, 3, 61-75.

[2] Narita, K *Trans Iron Steel Inst Jpn* 1975, 15, 145-152.

[3] Gladman, T In The Physical Metallurgy of Microalloyed Steels, The Institute of Materials, London, (1997).

[4] Vega, M I ; Medina, S F ; Quispe, A ; Gómez, M ; Gómez, P P *Mater Sci Eng A*, 2006, 423, 253-261.

[5] Gao N ; Baker T N (1998) *ISIJ Int* 1998, 38, 744-751.

[6] Medina S F ; Gómez M ; Rodríguez E ; Rancel L *ISIJ Int*, 2008, 48, 263-1269.

[7] Quispe A , Medina S F , Valles P *ISIJ Int*, 1997, 37, 783-788.

[8] Cabibbo M ; Fabrizi A ; Merlin M *J Mater Sci*, 2008, 43, 6857-6865.

[9] Mintz, B ; Arrowsmith, J M In Proc Int Conf on Hot Working and Forming Processes, ed by C M Sellars and G J Davies,Metal Society, London, 1980, 99-107.

[10] Funnell, G D In Proc Int Conf on Hot Working and Forming Processes, ed by C M Sellars and G J Davies, Metal Society, London , 1980, 104-107.

[11] Massardier, V ; Guétaz, V ; Merlin, J ; Soler, M *Mater Sci Eng A*, 2003, 355, 299-310.

[12] Sennour, M ; Esnouf, C *Acta Mater* 2003, 51, 943-957.

[13] Leap, M J ; Brown E L *Scripta Mater* 2002, 47, 793-797.

[14] Furuhara, T ; Yamaguchi, J ; Sugita, N ; Miyamoto, G ; Maki, T *ISIJ Int*, 2003, 43, 1630-1639.

[15] Hillert, M ; Staffanson, L I *Acta Chem Scand*, 1970, 24, 3618-3626.

[16] Medina, S F ; Hernández, C A *Acta Mater*, 1996, 44, 149-154.

[17] Andrade, H L ; Akben M G ; Jonas, J J *Metall Trans A*, 1993, 14A, 1967-1977.
[18] Medina, S F ; Gómez, M ; Chaves, J I ; Gómez, P P ; Adeva, P *Mater Sci Forum*, 2005, 500-501, 371-378.
[19] Orowan, E In Internal Stress in Metals and Alloys, The Institute of Metals, London, 1948, 451-463.
[20] Ashby, M F In Oxide dispersion Strengthening, ed by G S Ansell, T D Cooper and F V Lenel, Gordon and Breach, New York, 1958, 143-205.
[21] Le Bon, A ;Rofes-Vernis, J ; Rossard, C *Mèm Scient Rév Metal,* 1973, 55, 577-588.
[22] Le Bon, A ; J Rofes-Vemis, A J ; Rossard C *Met Sci*, 1975, 9, 36-40.
[23] Michel, J P ; M Akben, M ; Jonas, J J *Revue Metall*, 1981, 78, 823-833.
[24] Sellars, C M In Hot Working and Forming Processes, ed by C M Sellars and G J Davies, The Metals Society, London, 1980, 3-15.
[25] Sellars, C M *Czech J Phys B*, 1985, 35, 239-248.
[26] Beynon J H ; Sellars, C M *ISIJ Int,* 1992, 32, 359-367.
[27] Laasraoui, A ; Jonas, J J *ISIJ Int* 1991, 31, 95-105.
[28] Medina, S F ; Hernandez, C A *Mèm Scient Rev Mét*, 1992, 66, 217-223.
[29] Choi, J K ; Seo, D H ; Lee, J S ; Um, K K ; Choo, W Y *ISIJ Int*, 2003, 43, 746-754.
[30] Roberts, W ; Bodén, H ; Ahlbom, B ; Metal Sci, 1979, 13, 195-205.
[31] Samuel, F H ; Yue, S ; Jonas, J J ; Barnes, K R *ISIJ Int*, 1990, 30, 216-225.
[32] Pussegoda, L N ; Yue, S ; Jonas, J J *Metall Trans*, 1990, 21A, 153-164.
[33] Pussegoda, L N ; Jonas, J J *ISIJ Int*, 1991, 31, 278-288.
[34] Medina, S F ; Hernández, C *Acta Mater*, 1996, 44, 137-148
[35] Medina, S F ; Hernández, C A *Acta Mater,* 1996, 44, 149-154.
[36] Hernández, C A ; Medina, S F ; Ruiz, J *Acta Mater,*1996, 44, 155-163.
[37] Medina, S F ; Hernández, C A *Acta Mater,* 1996, 44, 165-171.
[38] Pussegoda L N ; Hodgson P D ; Jonas J J *Mater Sci Technol*, 1982, 8, 63-71.
[39] Sellars, C M *Mater Sci Technol*, 1990, 6, 1072-1081.
[40] Nanba, S ; Kitamura, M; Shimada, M; Katsumata, M ; Inque, T ; Imamura, H ; Maeda, Y ; Hattori, S *ISIJ Int*, 1992, 32, 377-386.
[41] Choquet, P ; Lamberterie, B ; Perdrix, C ; Biausser, H In Proc 4th Int Steel Rolling Conference, ed by IRSID, Deauville (France), 1987, B5 1.
[42] Siwecki, T *ISIJ Int*, 1992, 32, 368-376.
[43] Anelli, E *ISIJ Int*, 1992, 32, 440-450.
[44] Kwon, O *ISIJ Int*, 1992, 32, 350-358.
[45] Laarasoui, A ; Jonas J J *Metall Trans A*, 1991, 22, 1545-1547.
[46] Hodgson, P D ; Gibbs, R K *ISIJ Int*, 1992, 32, 1329-1338.
[47] Sun, W P ; Militzer, M ; Hawbolt, E B ; Meadowcroft, T R *Iron Steelmaker*, 1998, 12, 85-94.
[48] Yamamoto, S ; Ouchi, C ; T Osuka, T In Proc Int Conf on Thermomechanical Processing of Microalloyed Austenite, ed by A J De Ardo, G A Ratz and P J Wray The Metallurgical Society of AIME, Pensylvania, 1982, 613-639.
[49] Jonas, J J ; Akben, M G *Metals Forum*, 1981, 4, 92-101.
[50] Medina, S F; Mancilla, J E *ISIJ Int*, 1996, 36, 1063-1069.
[51] Medina, S F; Mancilla, J E *ISIJ Int*, 1996, 36, 1070-1076.
[52] Medina, S F; Quispe, A *ISIJ Int*, 2001, 41, 774-781.
[53] Medina, S F; Quispe, A ; Gómez M *Mater Sci Technol*, 2003, 19, 99-108.

[54] Toth, L E; A W Searcy, A W *Trans Met Soc AIME*, 1964, 230, 690-694.
[55] Kwon, O; DeArdo *Acta Metall Mater*, 1990, 39, 529-538.
[56] Quispe, A ; Medina, S F ; Cabrera, J M ; Prado, J M *Mater Sci Technol*, 1999, 15, 635-642.
[57] Medina, F; Quispe, A ; Valles, P ; Baños, J L *ISIJ Int*, 1999, 39, 913-922.
[58] Medina, F; Quispe, A ; Gómez, M :, *Mat Sci Technol*, 2001, 17, 536-544.
[59] Kang, K B; Kwon, O ; Lee, W B ; Park, C G *Scripta Mater*, 1997, 36, 1303-1308.
[60] Ardell, J *Acta Metall*, 1972, 20, 61-71
[61] Quispe, A ; Medina, S F ; Gómez, M ; Chaves, J I *Mater Sci Eng A,* 2007, 447, 11-18.
[62] Madariaga, I ; Gutiérrez, I *Acta Mater*, 1999, 47, 951-960.
[63] Shim, J H ; Oh, Y J ; Such, J Y ; Cho, Y W ; Shim, J D ; Byun, J S ; Lee, D N *Acta Mater*, 2001, 49, 2115-2122.
[64] Shim, J H ; Cho, Y W ; Cheng, S H ; Shim, J D ; Lee, D N *Acta Mater*, 1999, 47, 2751-2760.
[65] Byun, J S ; Shim, J H ; Cho, Y W ; Lee, D N *Acta Mater*, 2003, 51, 1593-1606.
[66] Al Hajari, K F ; Garcia, C I ; Hua, M ; DeArdo, A *ISIJ Int*, 2006, 46, 1233-1240.
[67] Furuhara, T; Shinyoshi, T ; Miyamoto, G ; Yamaguchi, J ; Sugita,N ; Kimura, N ; Takemura, N ; Maki, T *ISIJ Int*, 2003, 43, 2028-2037.
[68] Miyamoto, G ; Shinyoshi, T ; Yamaguchi, J ; Furuhara, T ; Maki, T ; Uemori, U *Scripta Mater*, 2003, 48, 371-377.
[69] Yokomizo, T ; Enomoto, M ; Umezawa, O ; Spanos, G ; Rosenberg, R O *Mater Sci Eng A*, 2003, 344, 261-267.
[70] Furuhara, T ; Yamaguchi, J ; Sugita, N ; Miyamoto, G ; Maki, T *ISIJ Int*, 2003, 43, 1630-1639.
[71] Medina, S F; Quispe, A *Mater Sci Forum*, 2003, 426-432, 1139-1144.
[72] Medina, S F ; Gómez, M ; Chaves, J I ; Gómez, P P ; Adeva, P *Mater Sci Forum*, 2005, 500-501, 371-378.
[73] Medina, S F; Quispe, A ; Gómez, M *Steel Res Int*, 2005, 76, 527-531.
[74] Liu, J ; Yanagimoto, J *ISIJ Int*, 2007, 47, 1188-1194.
[75] Medina, S F; Gómez, M ; Rancel, L *Scripta Mater,* 2008, 58, 1110-1113.
[76] Medina, S F ; Rancel, L ; Gómez, M ; Ishak, R ; De Sanctis, M *ISIJ Int*, 2008, 48, 1603-1608.
[77] Samuel, F H ; Yue, S ; Jonas, J J ; Zbinden, B A *ISIJ Int,* 1989, 29, 878-886.
[78] Najafi-Zadeh, A ; Yue, S ; Jonas, J J *ISIJ Int*, 1992, 32, 213-221.
[79] Karjalainen, L P; Maccagno, T M; Jonas, J J *ISIJ Int*, 1995, 35, 1523-1531.
[80] Kojima, A ; Watanabe, Y ; Terada, Y ; Yoshie, A ; Tamehiro, H *ISIJ Int*, 1996, 36, 603-610.
[81] Medina, S F ; Chapa, M ; Veja, M I *Mater Sci Technol*, 2000, 16, 163-170.
[82] Gómez, M ; Hernanz, O ; Medina, S F ; Tarín, P *Steel Res*, 2002, 73, 446-452.
[83] Medina, S F ; Gómez, M ; Rodríguez, E ; Rancel, L *ISIJ Int*, 2008, 48, 1263-1269.
[84] Gómez, M ; Medina, S F ; Quispe, A *Rev Metal Madrid*, 2005, Vol Extr, 379-383.
[85] Gao, N ; Baker, T N *ISIJ Int,* 1998, 38, 744-751.
[86] Rancel, L ; Gómez, M ; Medina, S F *Steel Res Int*, 2008, 79, 947-953.
[87] Rancel, L Bainitic transformations and improvement of mechanical properties in high strength steels Doctoral Thesis Madrid Complutense University, 2010.
[88] Manohar, P A ; Dunne, D P ; Chandra, T ; Killmore, C R ; *ISIJ Int*, 1996, 36, 194-200.

[89] Honeycombe, R W K; Pickering, F B *Metall Trans*, 1972, 3, 1099-1112.
[90] Bush, M E; Kelly, P M *Acta Metall*, 1971, 19, 1363-1371.
[91] Smith, D W; Hehemann, R F *J Iron Steel Inst*, 1971, 209, 476-481.
[92] Brozzo, P ; Buzzichelli, G; Mascanzoni, A ; Mirabile, M *Met Sci*, 1977, 11, 123-29.

In: Vanadium
Editors: V. N. Baranova and A. V. Fortunatov
ISBN 978-1-62081-600-4

Chapter 3

FERRITIC STAINLESS STEEL PROTECTED VANADIUM ALLOY AS CLADDING MATERIAL FOR FAST REACTORS

S. N. Votinov, V. P. Kolotushkin* and Y. V. Karasev
JSC "VNIINM " A. A. Bochvar, Moscow, Russia

ABSTRACT

The conditions of cladding structural materials work of promising fast reactors, as well as opportunities for the different classes materials use are analyzed. The paper demonstrates the advantages of vanadium base alloys with titanium, and chromium. Showing the principal foundations of structural materials technology for fuel rod claddings of fast reactors under conditions where corrosion resistance is provided by type Fe-17Cr ferritic stainless steel, which has the best corrosion resistance in atmospheric conditions and in pond, and the radiation resistance and heat resistance – by vanadium base alloys.

1. INTRODUCTION

The last 50 years have demonstrated that nuclear power engineering has no alternative fuel competition. Calorific value of uranium fuel exceeding the calorific value of fossil fuel equivalent of about 50 million times, ensures its competitiveness in comparison with fossil fuels, as well as the competitiveness of nuclear power in relation to "fire." This is due to the experimentally confirmed possibility of achieving in the fast-neutron reactors expanded breeding of nuclear fuel by conversion ^{238}U in ^{239}Pu and ^{232}Th in ^{233}U.

The strategic development of nuclear power requires to the arrangement of closed-cycle enterprises producing nuclear fuel and reprocessing the spent nuclear fuel, provided always engage into the fuel cycle of fast reactors (FR). Therefore, the creation of the effective industrial nuclear power plants with fast reactors (under construction sodium cooled BN-800

* E-mail: kolotush@bochvar.ru.

reactor and the projected lead cooled BREST-300) belongs to the priorities of the closure of nuclear fuel cycle.

Elevated requirements to the exploitation conditions of the active zones component parts of such reactors (operating temperature of fuel rod cladding over 750°C at the level of accumulation of radiation damage to more than 180 dpa) combined with increased demands for safety and the environment imposed to the acting settings.

An introduction to the nuclear power industry of new and advanced reactor FR requires increasing the service characteristics of the fuel elements. In turn, with the creation of advanced reactor FR fuel rods is very closely related problem of creating new structural materials that can satisfy stringent performance requirements.

When choosing of structural materials the principal demands are made as of the initial technological and physical-mechanical as well as to radiation and corrosion properties. Fuel elements in fast reactors operate at high energy intensity: neutron fluence ~ $2x10^{27}$ n/m^2 (E> 0.1 MeV), the temperature range along a fuel rod (370-710°C), stresses from the interaction of fuel with the shell and from gaseous fission fragments, corrosive interaction with liquid metal coolant (outside the shell), with the fuel and its fission products (in the shell).

Due to the stricter requirements on reliability, safety and environmental characteristics of new generation fast reactors, a significant increase operational parameters (temperature, fluence, thermal-stress, resource) materials of austenitic and ferritic-martensitic (FM) class developed to date can not provide reliable operation of the fuel elements of advanced nuclear plants [1].

2. Comparative Evaluation of Existing Structural Cladding Materials

Many years of experience in domestic and foreign reactor constructions showed that the success of the creation of structural materials for fuel cladding of reactor FR is achieved while maintaining high level of three major factors :

1. The complex of physical-mechanical and technological properties (including weldability).
2. Radiation resistance.
3. Compatibility with the coolant and the fuel material.

The most important criterion in the evaluation of structural materials for the conditions under consideration is the heat resistance, because it determines in designing the installation conformity of the material requirements, including high temperature and applied mechanical stress.

2.1. Austenitic Steels

At one time the most preferred fuel rod cladding materials of reactor FR have been the austenitic steels due to a satisfactory combination of service properties: strength, heat resistance, corrosion resistance, good process ability and development of industrial [2]. During long-term tests austenitic steels have an advantage also before the high nickel alloys. Up to 1000 hours high nickel alloys have a higher long-term strength, and then due to the intensive coagulation of the hardening phase Ni_3 (Ti, Al), they become less strong. The limit of long-term strength at 700°C, determined by extrapolation to 100 thousand hours, for high nickel alloy EI 437 is 80 MPa, and for austenitic steel of 140-150 MPa [3].

However, the propensity of austenitic steels to high-temperature radiation embrittlement and, more importantly, to the vacancy swelling significantly reduces the possibility of achieving the necessary high nuclear fuel burnup.

2.2. Chromium Ferritic-Martensitic Steels

Steels in this class well recommended themselves when the temperature of the plants does not exceed 550°C [2]. The main disadvantages of 12% chromium steels are the low heat resistance at temperatures above 550-600°C, the tendency to low-temperature radiation embrittlement and the significant loss of strength during the neutron irradiation at temperatures above 550°C. In addition, at temperatures up to 550°C, these steels also have a tendency to swelling. Reducing the heat resistance of chromium steels at temperatures higher than 550°C is associated with coagulation of carbides on the subgrain boundaries of sorbite and annealing of dislocations.

Summarizing a brief overview, we can emphasize that in connection with the toughening of requirements on reliability, safety and the environmental characteristics in the new-generation reactors and substantial increasing the operating parameters of fuel rod cladding (working temperature of 320-710°C, achieved damaging doses of 140-180 dpa, the district stresses up to 80-120 MPa, an increase of the resource to 5-7 years) austenitic and ferritic-martensitic class steels, developed to date can not provide reliable operation of fuel rods of perspective fast reactors.

Some time back as a perspective structural material for the fuel claddings of fast neutron reactors were considered alloys of vanadium. However, experiments in England and the Soviet Union have shown full unsuitability of vanadium alloys in a sodium coolant [4, 5].

Nevertheless, interest in vanadium alloys was reborn in connection with the problems of fusion reactor [1, 6, 7].

3. Vanadium Base Alloys

3.1. The Activation Properties of V-Ti-Cr

It is known that vanadium is more than the others structural materials corresponds to the criterion of small activation [8]. Alloying of vanadium with titanium and chromium does not output the alloys from the category of little activated, with the biggest problem may represent titanium isotopes ^{46}Ti and ^{47}Ti. The remaining isotopes of titanium does not give long-lived radio nuclides.

Activation characteristics of vanadium alloys deteriorate in the presence of impurities Ni, Nb, Mo, Co, Ag, Bi, Cu, Sm, etc. Calculation of activation parameters of vanadium alloys irradiated in the position of the first wall of DEMO to a fluence of 35 MW-year/m^2 (12 years of continuous irradiation at middle neutron load on the first wall 2.91 MW/m^2) [9] shows that within acceptable to the technically holding periods of 10-40 years, the main contribution (> 90%) in the dose rate of the alloy V-5Ti-5Cr without impurities makes isotope ^{42}K, and of the alloy V-5Ti-5Cr with activated impurities - ^{94}Nb (table 1).

Table 1.The activation characteristics of V-5Ti-5Cr (12 years exposure at an average neutron load 2.91 MW/m^2)* on the first wall [9]

Characteristics	Exposure time, years					
	1	5	10	20	40	100
The residual energy release, W / kg	$1 \cdot 10^{-1}$ ($1 \cdot 10^{-1}$)	$5 \cdot 10^{-4}$ ($6 \cdot 10^{-3}$)	$3 \cdot 10^{-5}$ ($3 \cdot 10^{-3}$)	$1 \cdot 10^{-5}$ ($7 \cdot 10^{-4}$)	$7 \cdot 10^{-6}$ ($7 \cdot 10^{-5}$)	$2 \cdot 10^{-6}$ ($7 \cdot 10^{-6}$)
Contact dose rate of γ - radiation (micro-Sievert / h)	$1{,}1 \cdot 10^{8}$ ($1{,}2 \cdot 10^{8}$)	$3 \cdot 10^{4}$ ($8 \cdot 10^{6}$)	$4 \cdot 10^{3}$ ($5 \cdot 10^{6}$)	$3{,}3 \cdot 10^{3}$ ($1{,}2 \cdot 10^{6}$)	$2{,}1 \cdot 10^{3}$ ($1 \cdot 10^{5}$)	$5 \cdot 10^{2}$ ($4 \cdot 10^{3}$)
Radio nuclides, which provide the main contribution to the dose rate of the alloy	^{46}Sc, ^{51}Cr, ^{54}Mn (^{46}Sc, ^{60}Co, ^{54}Mn)	^{42}K, ^{54}Mn (^{182}Ta, ^{60}Co, ^{54}Mn)	^{42}K (^{60}Co)	^{42}K (^{60}Co, ^{154}Eu)	^{42}K (^{154}Eu, ^{94}Nb)	^{42}K (^{94}Nb)

* The first value - for the alloy without the activated impurities, the second (in brackets) - for the alloy with activated impurities.

Calculated estimates of designs activity of one of the variants of DEMO reactor have shown [9] that, when used as the construction materials of the reactor of conventional or modified austenitic and ferritic steels, the vanadium alloys produced by industry, as well as copper alloys (such as brazing material) and the superconductor Nb_3Sn the induced activity will be quite high. This could severely complicate the maintenance of the reactor during its operation, and dismantling, re-fabrication of the irradiated materials and disposal of waste.

In this connection are preferable, no doubt, alloys based on vanadium. However, realization of their potentials encounters many difficulties.

The presence of a natural mixture of isotopes of titanium in the alloys results in the occurrence of a significant amount of the isotope ^{42}Ar: its decomposition with the formation and subsequent decay of ^{42}K results in a period of 20-100 years or more after stopping the reactor to increase by about two orders of magnitude the dose rate γ - radiation of alloy with titanium compared with the alloy without titanium.

An even greater degree worsens the situation technological impurities - cobalt, nickel, copper, iron, silver, europium, niobium and others.

Workable alloy based on vanadium for fusion energy, you can manually or using the minimum of personal protective equipment to carry out the dismantling the reactor and the re-fabrication of the irradiated alloy must have a deep degree of purification from impurities such as Co, Ni, Cu, Ag, Eu, which modern technology to provide can not [8].

However it is necessary to emphasize the fact that by using alloys based on vanadium as the structural materials for fast reactor the activation problems and activation characteristics of alloys do not have such fundamental importance as their use for fusion reactors.

3.2. The Effect of Doping on the Physical and Mechanical Properties of Vanadium

With many elements (Ti, Cr, Ta, W, Fe, Nb, Mo) vanadium forms the continuous series of solid solutions or wide areas of solubility. Some of the elements (Si, C, Sc) have limited solubility in vanadium (Table 2 [8]). These circumstances allow to implement both mechanisms of alloys hardening —solid solution and dispersion, which opens up the broad opportunities for the development of vanadium alloys with the required level of physical and mechanical properties [10-12]. However, the requirements to the fusion reactor structural materials for limiting the activation significantly restrict the list. Cr and Ti belong to the main alloying elements, which support the weak activation of alloys. Alloyed with them alloys have strength of 400-500 MPa, and at concentrations above 5 at% it varies only slightly (Figure 1) [13].

Table 2. The solubility of elements in vanadium at 1000°C (%)

Element	Solubility	Element	Solubility	Element	Solubility
Nb, Cr, Mo, W	100	Al, Au	28	Hf	2
Ti	95	Pt	25	Si	0,5—2,5
Re	85	Sn	18	Be, N	1
Mn	52	Ga	12—13	Mg	0,8
Fe, Ir, Os	36—38	Co	8	B	0,43
Ta	27—35	Ni	7	O	0,1
Ru	35	Cu	7,5	C	0,1
Pd	30	Zr	6		

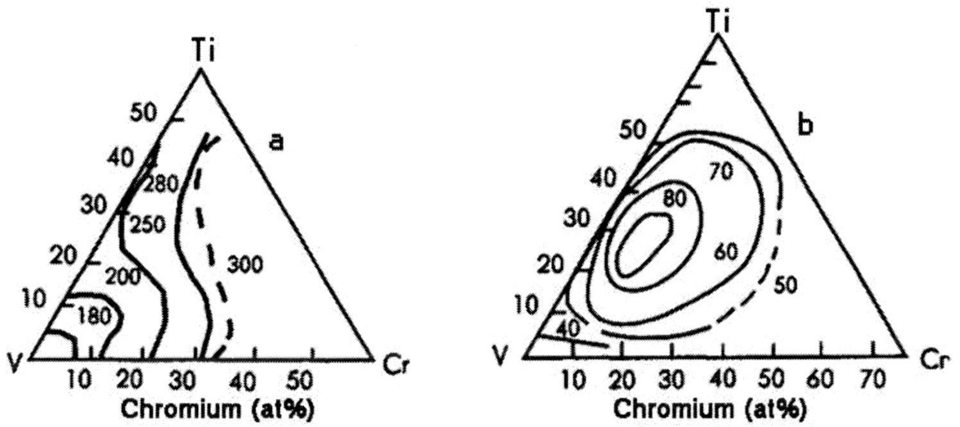

Figure 1. The curves of similar micro hardness values (a) and the limit of long-term strength at 600°C (b) for alloys of vanadium-titanium-chromium kg/mm^2 [13].

Metallic impurities contained in the original vanadium, have virtually no effect on the hardness, strength properties and grain size of vanadium-titanium alloys. It is found in the study of vanadium grades VNM-1 and VE with different contents of metallic impurities (Table 3 [8]). Constraining factors using the vanadium-based alloys considered to be their embrittlement during the interaction with oxygen (Table 4), whose solubility in vanadium at operating temperatures high enough [10, 14, 15].

Table 3. The content of impurities in vanadium, wt%

Brand of vanadium	Elements								
	Al	Si	Fe	Ni	S	C	N	O	H
VNM-1	0.2	<0.2	0.07	—	—	0.017	<0.01	0.015	0.001
VE	<0.01	0.01	0.02	<0.01	< 0.005	<0.01	<0.01	0.05	—

The temperature dependence of the brittle-ductile transition on the amount of impurities, which determines the permissible limits of interstitial impurities in vanadium, is shown in Figure 2. The temperature of the brittle-ductile transition of 0°C corresponds to the content of impurities: H - 0.015, O - 0.15, N - 0.22 at%. These impurities abruptly alter the temperature of the brittle-ductile transition. Carbon because of its low solubility is almost no effect.

Table 4. The solubility of interstitial impurities in metals (at%)

Metal	Interstitial impurities						
	Carbon			Nitrogen		Oxygen	
	400°C	1000°C	1600°C	400°C	1600°C	400°C	1500°C
V	–	$2,5 \cdot 10^{-3}$	3,7	1,8	11,0	2,3	8,0
Nb	–	$3 \cdot 10^{-3}$	0,39	$5 \cdot 10^{-2}$	1,0	0,13	5,2
Fe	$5 \cdot 10^{-2}$	6,4	–	0,1	–	10^{-6}	–
Cr	–	$6 \cdot 10^{-2}$	2,0	$3 \cdot 10^{-5}$	10^{-2}	10^{-6}	0,07

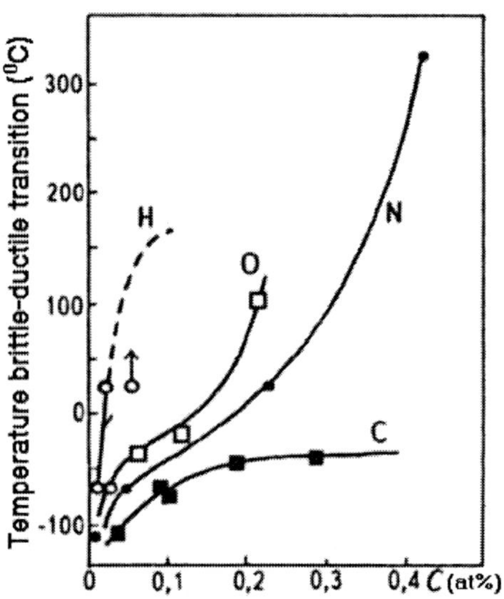

Figure 2. Effect of interstitial impurities on the ductile - brittle transition temperature of vanadium according to the bending tests [16].

Experimental results have shown that the vanadium alloys with additions of titanium 0 - 20% have a minimum value of the transition temperature to the brittle state. When the content of titanium is 5%, the transition temperature to the brittle state is less than - 196°C, which follows from the form of samples after impact bending tests (Figure 3 [17]). Chromium supplements up to 15% of the alloy significantly increase the value of the transition temperature to the brittle state (on the 90-160°C). Silicon additives (0.25-1) % to the alloy V-3Ti lead to an increase the transition temperature by 60°C. The presence of hydrogen (0.04-0.12) % increases the transition temperature on the 60-250°C.

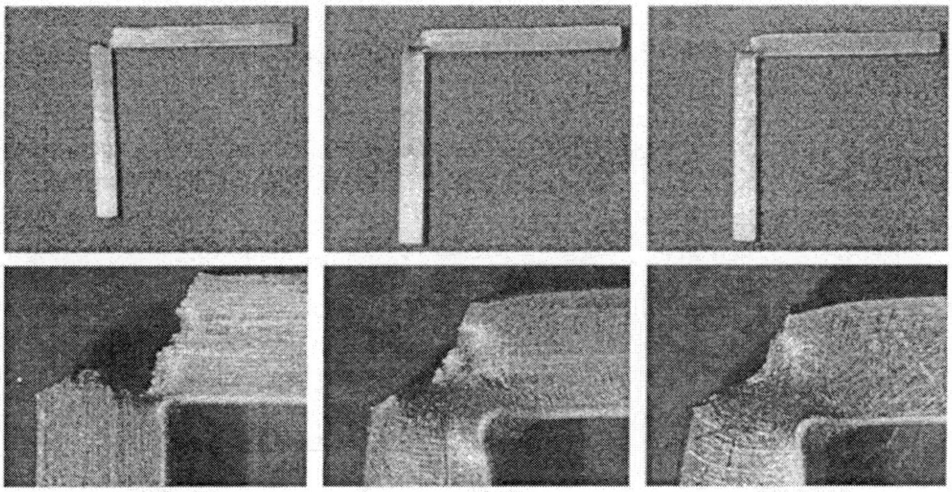

Figure 3. Samples of V-5Ti alloy after the impact test at temperatures: - 196°C, - 50°C and 100°C [17].

Table 5. The content of interstitial impurities in vanadium alloys at technological updating stages (at%) [18]

Type of intermediate product	Stage of technological process			
	Impurity	The initial mix material (total)	An ingot of the alloy	Semi-finished
Rods Ø 15 (mm)	Oxygen Nitrogen Carbon	0,03 0,007 0,03	0,04 0,007 0,025	0,06 0,007 0,03
Strip 1x500x100 (mm)	Oxygen Nitrogen Carbon	0,03 0,007 0,03	0,04 0,01 0,03	0,04 0,01 0,03
Sheet 5x500x1000 (mm)	Oxygen Nitrogen Carbon	0,05 0,007 0,025	0,04 0,007 0,025	0,06 0,007 0,03
Pipe Ø 16x1 (mm)	Oxygen Nitrogen Carbon	0,05 0,007 0,015	0,06 0,013 0,025	0,07 0,015 0,035

Also known that when the total impurity content is ~ 0.15 wt%, the temperature of brittle-ductile transition is ~ -125°C, and our investigations showed that the commercial brands vanadium VNM-0 and VNM-1 is quite suitable as a basis for the production of the technological vanadium base alloys V-Ti-Cr.

When analyzing of the vanadium alloys state it should also take into account that the impurities amount in the metal changes in technological stages of semis (Table 5).

In regard to the effect of impurities from the standpoint of manufacturability to the alloys based on vanadium are presented increased requirements, namely:

1. High purity of raw materials (by interstitial impurities and the activated);
2. Using of technological environments ensuring the protection from gas saturation at high temperature treatments.

In Figure 4 are compared the temperature dependences of the tensile strength and the total elongation of samples V-4Ti-4Cr and V-10Ti-5Cr at initial state and with a transverse weld. For the V-4Ti-4Cr are presented the data on the mechanical properties of specimens with welded joints produced by the two methods of welding: tungsten-arc inert-gas welding (TIG) and electron-beam welding (EBW). The presence of the weld on both samples does not change the nature of the curves. The ultimate strength of samples V-4Ti-4Sr with the weld a little depends on the kind of welding and the ultimate strength comparable to that the initial samples.

TIG also does not alter the strength of the samples V-10Ti-5Cr compared with initial state. However, the presence of welded joints made by electron-beam welding methods and in particular TIG, causes a significant reduction in ductility of the samples V-4Ti-4Cr compared with initial state. Reduction of the value of total elongation ($\Delta\delta_{tot}$) of TIG - samples of this alloy is 11.0-14.5% and 7.5-8.0% at the test temperature 20-200°C and 250-650°C, respectively. At all test temperatures is observed advantage of the method EBW.

Curve of the dependence $\delta_{com} \sim f(T_{test})$ for EBW - samples of the alloy V-4Ti-4Cr occupies an intermediate position between the original samples and the TIG - samples.

Under identical experimental conditions, reducing of δ_{com} as a result of welding for specimens V-10Ti-5Cr at the test temperature 20-200°C and 250-650°C is 11.0-12.0% and 6.0-7.5% respectively. At 700°C this effect is smallest - less than 3% regardless of the type of vanadium alloys and welding method [8].

Studies of recent years in connection with the fusion reactor is shown the perspective of using in plants with increased operating temperature and neutron fluence of vanadium base alloys. These alloys are most suitable materials [4. 6. 12. 13. 20. 21] in installations for nuclear and thermonuclear reactors. They have good nuclear-physical properties, such as a rapid decrease of induced radioactivity and its low level, radiation resistance and high long-term strength at 750°C (Table 6 [3]).

Meanwhile the FM class steels are softened at temperatures above 600°C and austenitic steel – 650°C. Vanadium base alloys (plus titanium, and chromium) have in the temperature range 550-700°C high indicators of the time to failure (Figure 5, the data of S.N. Votinov and V.P. Voeikov) [3]. The parametric dependence of the long durability of vanadium alloys (Figure 6) [22] demonstrates their higher properties compared to steel.

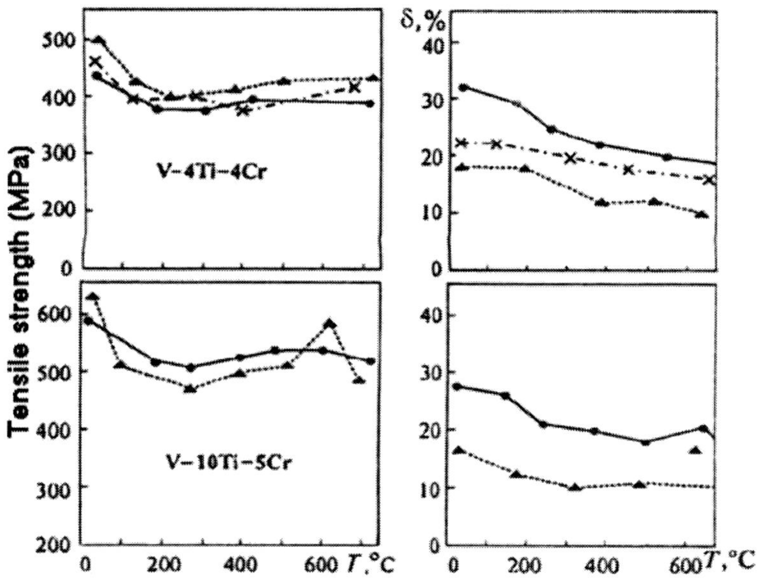

Figure 4. Temperature dependence of the strength characteristics and the relative elongation of samples of vanadium alloy weld: ● - basic metal; ▲ - TIG, x - EBW [19].

As noted above, the high strength properties at high temperature are considered to be one of the advantages of V-Ti-Cr alloys in comparison with other reactor materials. The vanadium alloys have a yield stress equal to 400 MPa at 800°C compared to austenitic and ferritic-martensitic steels, which soften at above 600°C and above 500°C, respectively. The creep rupture advantages of vanadium alloys are even greater [11, 23]. At 700°C for 10 000 h the long-term rupture strength of V-4Ti-4Cr and V-10Ti-5Cr alloys amount to 140 and 180 MPa, respectively.

Table 6. High-temperature properties of ferritic-martensitic, austenitic steels and vanadium alloys

The material, Content of elements	Long-term strength σ_{lts} on the basis of 10,000 hr depending on the temperature				
	Temperature (°C)				
	550	600	650	700	750
	Long-term strength (MPa)				
Ferritic-martensitic steels: 12 % Cr, 1 % Mo, W, V	300	180	90	< 50	–
Austenitic steels: 16 % Cr, 15 % Ni, 0,5 % Nb, 0,004 B	200	190	180	150	–
Alloys of V-Ti-Cr:					
92 % V, 4 % Ti, 4 % Cr	370	230	180	140	100
85 % V, 10 % Ti, 5 % Cr	–	300	–	180	–

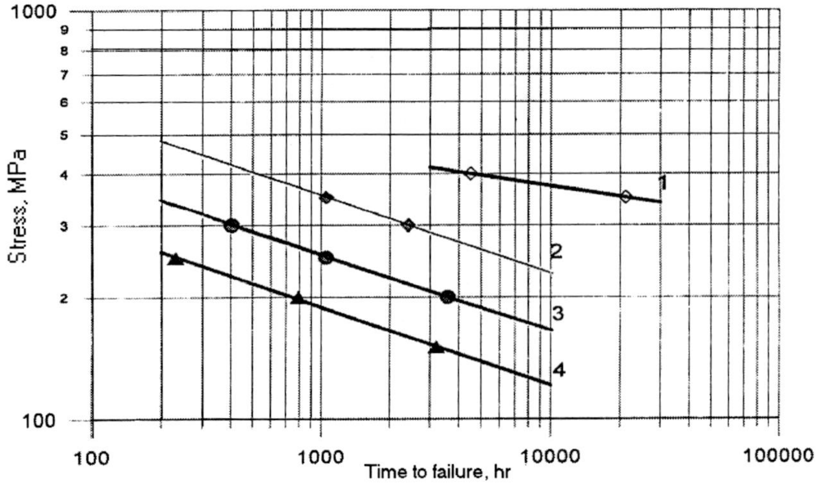

Figure 5. The stress at different temperatures: 1 – 550°C, 2 – 600°C, 3 – 650°C, 4 – 700°C vs. time to failure of V-4Ti-4Cr samples [3].

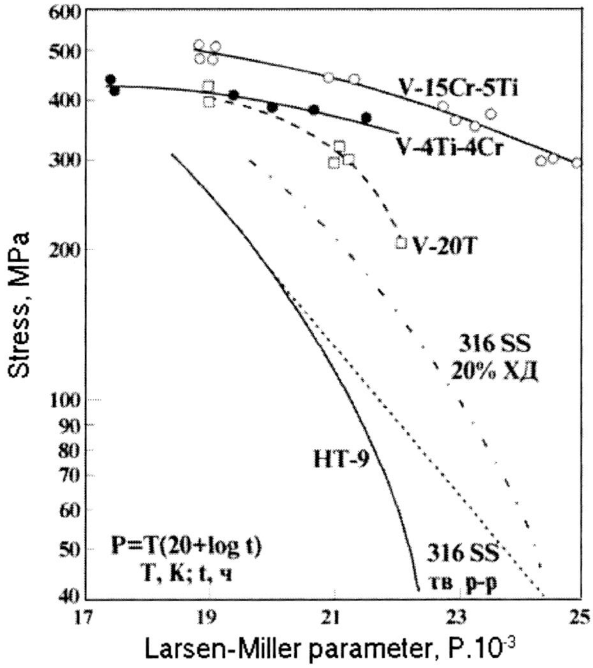

Figure 6. The parametric dependence of the long-term strength of vanadium alloys in comparison with steels [22].

At 600°C these values are 230 and 300 MPa, respectively. This should be particularly emphasized, since under certain conditions the short-term strength cannot be considered as a criterion in evaluated materials. This is illustrated in Figure 7, where the temperature dependence of short-term and long-term properties of V-Ti-4Cr alloy specimens is given [5].

At temperatures above 600°C it is useless to apply data on alloy short-term properties. However, taking into account the fact that the increase of yield stress correlate with a decrease in creep rate [24], then strengthening is indicative of improvements in creep rupture behavior. Since the value of $\sigma_{0.2}$ for V-10Ti-5Cr alloy is superior by about 30% over that for V-4Ti-4Cr, it is believed that there will be a marked increase in creep rupture resistance of the first alloy compared to the second.

Analysis and generalization of the available data on the properties of vanadium base alloys, V-4Ti-4Cr and V-10Ti-5Cr predicts that the maximum operating temperature of the first alloy can reach 750°C, and the second 750-800°C (Figure 8 [17]).

Mechanical properties of V-4Ti-4Cr and V-10Ti-5Cr alloys

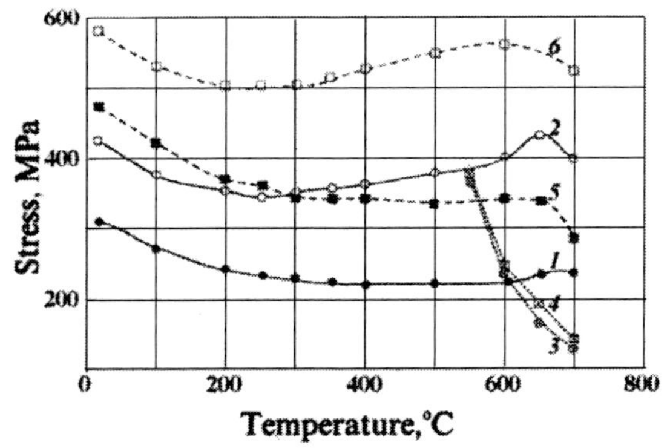

V-4Ti-4Cr 1- Yield stress, 2-ultimate strength,
3- long-term rupture strength (10000 h),
4- creep strength (ε=1 10–4 %/h),
V-10Ti-5Cr 5- Yield stress, 6- ultimate strength.

Figure 7. Mechanical properties of the vanadium alloys [5].

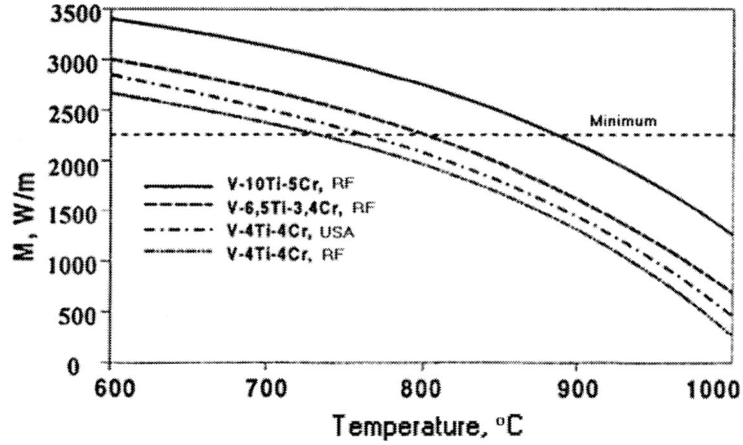

Figure 8. Determination of the maximum operating temperature using thermal strength parameter, $M=1,6[(1-\nu)\lambda\sigma_{\tau}] / (\alpha E)$ [17].

Table 7. Mechanical properties of the V-Ti-Cr alloys at different strain rates

Alloy	The test temperature (°C)	Strain rate (mm/min)	Ultimate strength (MPa)	Yield strength (MPa)	δ (%)
V-10Ti-5Cr	250	0,1	490	310	28
1075-30', vacuum		1	510	310	21
O – 0,040	500	0,1	520	320	24
C – 0,015		1	550	340	20
N – <0,01		10	520	330	25
V-4Ti-5Cr	250	0,1	310	200	29
1075-30', vacuum		1	350	240	26
O – 0,03	500	0,1	390	220	27
C – 0,01		1	380	225	20
N – <0,01		10	330	145	25

The advantages of the alloy V-10Ti-5Cr over the V-4Ti-4Cr are also manifested when comparing the mechanical properties determined after the tests at various rates of deformation. The data in Table 7 (and the analysis of deformation curves) demonstrate that, due to strain aging, mechanical properties of the alloy V-10Ti-5Cr significantly higher and more stable than those of the alloy V-4Ti-4Cr [8].

The alloys of vanadium containing 5-10 at% Ti and ~ 6 at% Cr, at 600°C have a modulus of elasticity E = (122 -124) GPa and shear G = (45 - 46) GPa. The elastic moduli of these composition alloys are 1.5 times lower than that of the austenitic stainless steels. The thermal conductivity of these alloys at 600°C is 29-31 W / (m · K), which is 30% higher than that of the austenitic steels (Table 8 [22]).

Table 8. The elastic and thermo physical properties of construction materials

Material	Temperature (°C)	λ (W / m · K)	$\alpha \cdot 10^6$ (K^{-1})	E (GPa)	μ
Ferritic steels	600	27-32	12	150-185	0,25–0,35
Austenitic steels	300	20	17	180	0,35
V	600	34	10,7	117,6	0,36
V–4Ti–4Cr	600	33	10	125	0,37
V–10Ti–5Cr	600	30	10	120	0,32

The designations: λ - thermal conductivity, α - coefficient of thermal expansion, E - elastic modulus, μ - Poisson's ratio.

3.3. Radiation Resistance of Alloys Based on Vanadium

The results of tests a series of alloys based on vanadium in the BOR-60 and BN-600 reactors have shown that the alloying with titanium decreases the degree of radiation hardening the vanadium, and the greater the higher the titanium content in the alloy. However alloys with the concentration of titanium more than 25% showed a propensity to high-temperature radiation embrittlement; reduction of high-temperature radiation embrittlement effect is reached at alloying the system V-Ti with chromium, but the high-chromium alloys are becoming inclined to the low-temperature radiation embrittlement.

In such a way the vanadium alloys can be inclined to the low temperature as well as to the high-temperature radiation embrittlement. In this case to the high-temperature radiation embrittlement more prone alloys with the predominant and significant (> 25-30%) alloying with titanium, and to the low-temperature radiation embrittlement - alloys alloyed with chromium mainly. The optimal must be considered alloying of vanadium with titanium and chromium with the total amount of 10-15% and the ratio of the content of titanium / chromium = 2 / 1 [8, 13]. The irradiation of such composition alloys in the temperature range 400-700°C does not lead to a substantial change in the strength properties and relative elongation.

3.3.1. Features of Structural Transformations in Alloys of Vanadium

The behavior of the vanadium alloys under irradiation can be connected with the structural changes. The phase diagram of the V-Ti system (Figure 9 [25]) enables to assume the dependence of the structural changes in alloys during neutron irradiation from the disintegration incubation phase of the solid solution on the α - Ti and titanium-rich vanadium. In the investigation of aging of the vanadium alloys with the 50 and 70% Ti [26]

X-ray analysis has shown that at 500°C in the alloys takes place transformation $\beta \rightarrow \alpha + \beta_1$, where the β - the original solid solution, and β_1 - solid solution enriched in vanadium. However, when the taken annealing conditions to achieve the completion of the transformation did not last and the diffraction patterns contained lines of all three phases. It can be assumed that the structural changes in alloys with lower content of titanium (<~ 30%), seems to stop at the stage of formation of short-range order.

In all transformations in the structure of transition metals alloys, which we observe in the aging, it is clearly can be seen caused by the desire of the system to the minimum the free energy the tendency to the site exchange of atoms solid solutions base with the formation of pairs of these atoms and the formation at homologous temperatures of short-range order. This trend is apparent, for example, when evaluating the micro hardness of the aged at temperatures of 350-450°C for 40 thousand hours alloys Ni-Cr-Mo, in which, in accordance with increasing duration of aging and extent of short-range order increases the micro hardness [27]. In alloys based on vanadium at appropriate their alloying similar this tendency is also clearly visible. For example, in [28] investigated the change of the hardness of cold-rolled alloys V-6W-4Ti, V-4Ti and V-4Cr-4Ti as after recrystallization at 900°C, well as after homogenization at 1100°C, plus annealing or aging. The results of the determination of hardness have shown that the particle growth of the excess phase was controlled by diffusion of interstitial atoms C, N and O. At the same time under identical conditions, alloy V-4Cr-4Ti showed much stronger precipitation hardening than the other two alloys due to the addition of Cr (Figure 10). The difference in the hardness changes of the homogenized and recrystallized alloys is due to, apparently, the fact that after recrystallization annealing the atoms W, Ti and Cr practically entirely connected to the large discharge particles.

The analysis shows that such a significant increase in hardness by the addition of chromium can not be due only to the formation of compounds such as Cr - (O, N, C), since the binding energy of chromium with these interstitial elements is less than of titanium or tungsten (eg. the dissociation energy of molecules MO is 663, 626 and 435 kJ / mol for Ti, V and Cr, respectively) [29].

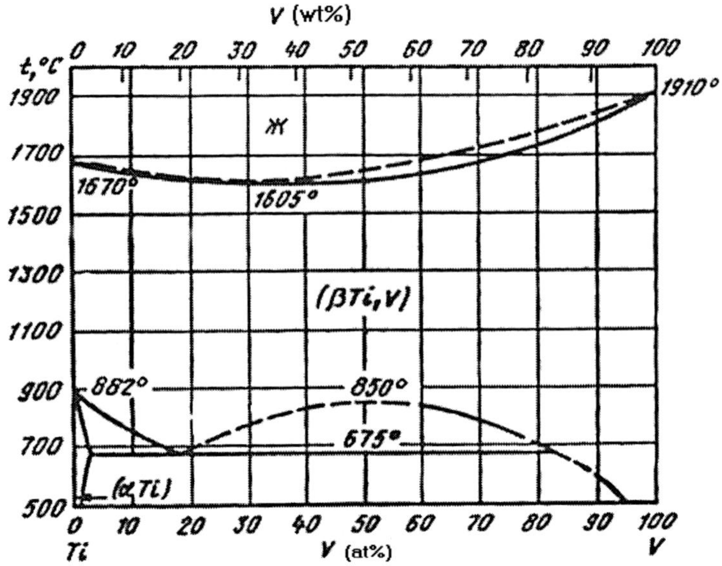

Figure 9. Ti-V phase diagram.

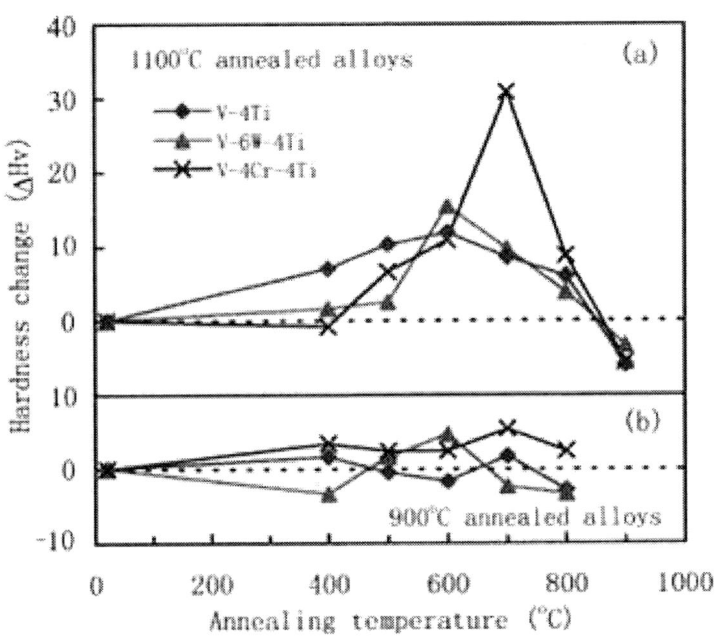

Figure 10. Hardness change as a function of annealing temperature for (a) the 1100°C annealed alloys and (b) the 900°C annealed alloys [28].

It follows from this fact that for the increase in hardness during annealing, along with the influence of excess phase particles, should be responsible to some other mechanism.

Such a mechanism could be, apparently, short-range ordering in alloys of the vanadium with titanium and chromium with the emergence of compound $TiCr_2$ (Figure 11 [30]).

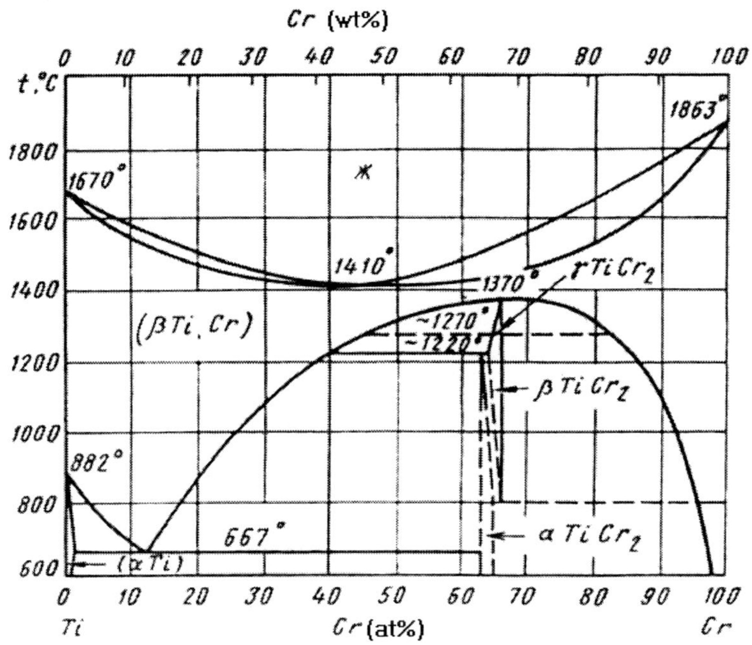

Figure 11. Ti-Cr phase diagram.

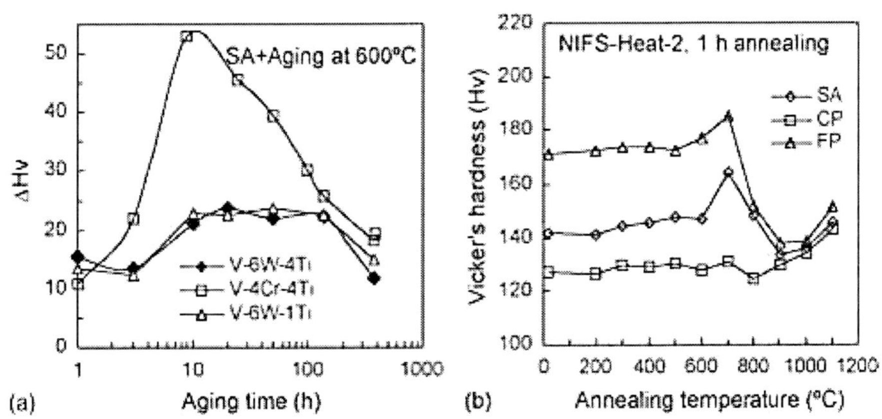

Figure 12. Vicker's hardness increment (ΔHv) as a function of the aging time for V–6W–4Ti, V–4Cr–4Ti (SWIP-Heat) and V–6W–1Ti alloys (a) and the dependence of the hardness on annealing temperature for the NIFS-Heat 2 alloy after SA, CP and FP treatment (b). Hardness was measured at room temperature [31].

The authors of [31] note that the additional hardening of the alloy V-4Cr-4Ti appears at a temperature of 600-700°C for homogenized (1100°C / 1 h), and for the aged (1100°C / 1 h + 600°C/10 h) samples. With an increase in annealing temperature to 900°C, the hardness decreases and then increases at 1000 -1100°C. At temperatures below 500°C the hardness of aged (1100°C / 1 h + 600°C/10 h) alloy V-4Cr-4Ti is quite stable (Figure 12).

Concurrent testing of the alloy V-4Cr-4Ti (SWIP-Heat) samples in tension at room temperature (Figure 13 [31]) showed that following the elastic regime the yielding and Luders expansion occur only in the homogenized alloy. Instead, the alloy after the aging shows an immediate hardening. On the other hand, the aged alloy has a higher yield strength and tensile strength than non-aged alloy. Such results are, in our opinion, due to a tendency of the alloy V-4Cr-4Ti to the formation of short-range order at temperatures of 600 ± 100°C.

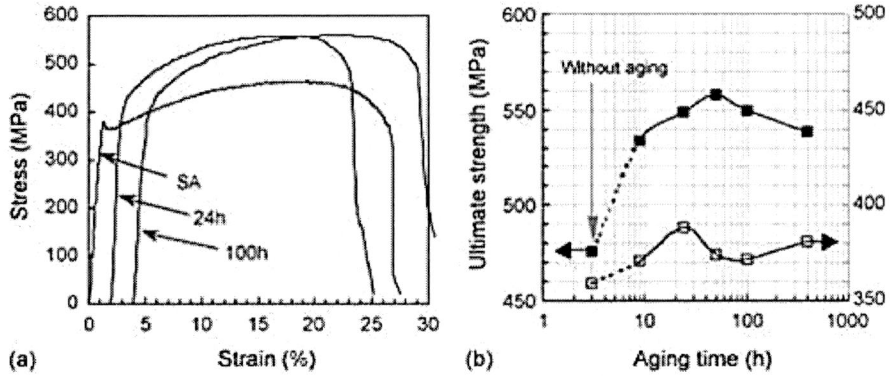

Figure 13. Tensile curves for SA treated V–4Cr–4Ti (SWIP-Heat) specimen and those after 24 and 100 h aging (a) and tensile strengths of the SA alloy as a function of the aging time at 600°C (b). All tested at room temperature [31].

Changing the amount of oxygen and nitrogen impurities in the alloys is also influences the evolution of the defect structure during neutron irradiation as well as on the occurrence of phase and segregation phenomena. According to the phase diagram [3] for the ternary V-Ti-O system (Figure 14, the diagram is similar to nitrogen), the oxygen solubility decreases sharply with increasing Ti content and increases with temperature.

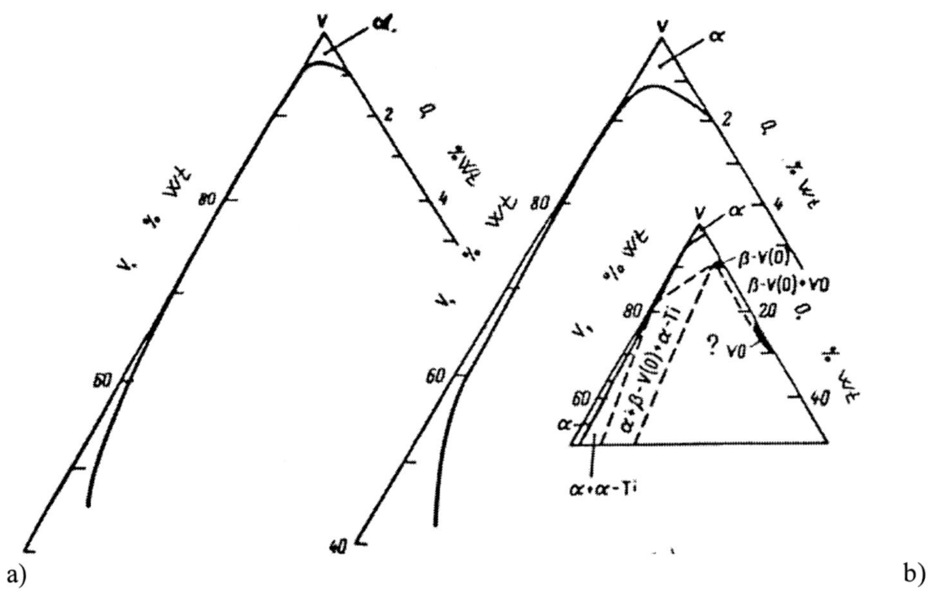

Figure 14. Phase equilibriums in the vanadium corner of the ternary system V-Ti-O: a - at 800°C, b - at 1200°C.

The results of experiments on the irradiation with neutrons indicate extremely little change in mechanical properties (including the temperature of the brittle-ductile transition T_{br}) of the vanadium alloys at 420-600°C and above. The irradiation at temperatures below 400°C results in more substantial change in properties of alloys (Figure 15 [17]). However features of embrittlement of the vanadium alloys are not observed.

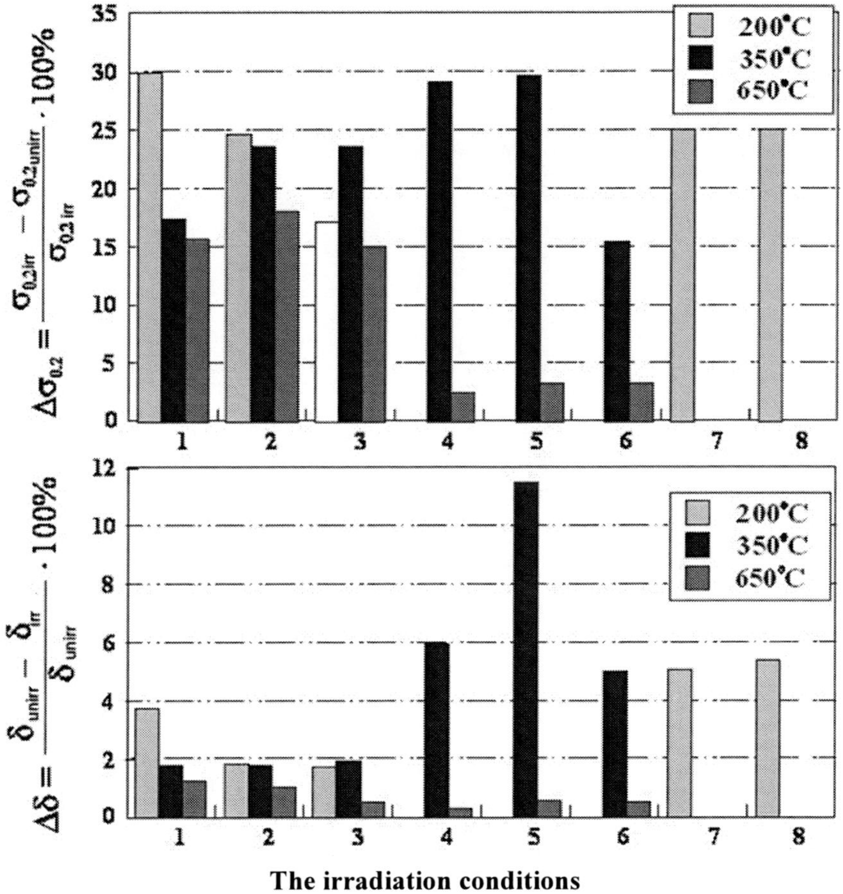

№	Alloy	Nuclear reactor	Irradiation temperature (°C)	Fluence
1	Vanadium	BOR-60	780	43 dpa
2	V-4Ti-4Cr	BOR-60	450	22 dpa
3	V-10Ti-5Cr	BOR-60	450	22 dpa
4	Vanadium	BOR-60	330	18 dpa
5	V-4Ti-4Cr	BOR-60	330	18 dpa
6	V-10Ti-5Cr	BOR-60	330	18 dpa
7	V-4Ti-4Cr	IVV-2 M	-196	$1.5 \cdot 10^{22}$ n/m^2
8	V-4Ti-4Cr	IVV-2 M	-196	$1.5 \cdot 10^{22}$ n/m^2

Figure 15. Change in yield strength and elongation of vanadium, V-4Ti-4Cr and V-10Ti-5Cr as a function of irradiation and test temperature [17].

Radiation swelling of vanadium and its alloys, as shown by experimental data, can vary within very broad limits [32]. High-purity vanadium with the stable structural-phase state is subject to the intensive swelling (Figure 16).

Alloying with elements with the size of an atom smaller than that of vanadium (iron, chromium [33]), increases the capture of interstitials by dislocation loops and, consequently, the swelling [34] (Figure 17 [3]), and the addition of elements, which increase the lattice constant of vanadium (tungsten, molybdenum, titanium), reduce the swelling (Figure 18).

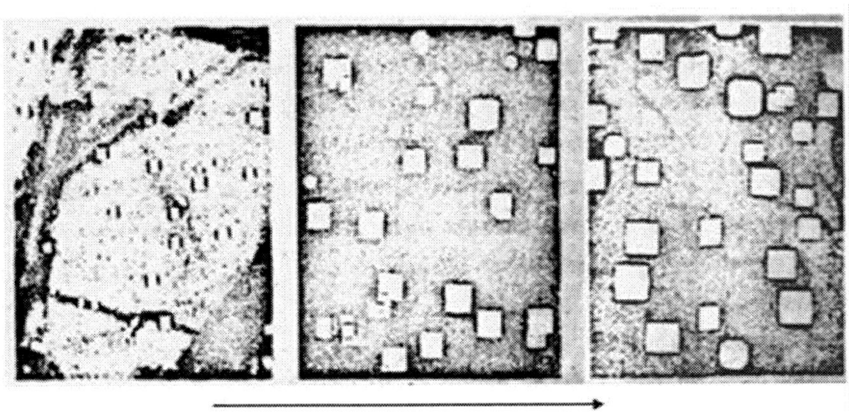

Figure 16. Micrographs of voids in high purity vanadium, irradiated by vanadium ions with an energy of 3 MeV at 700°C (arrow indicates the increase of the dose) [32].

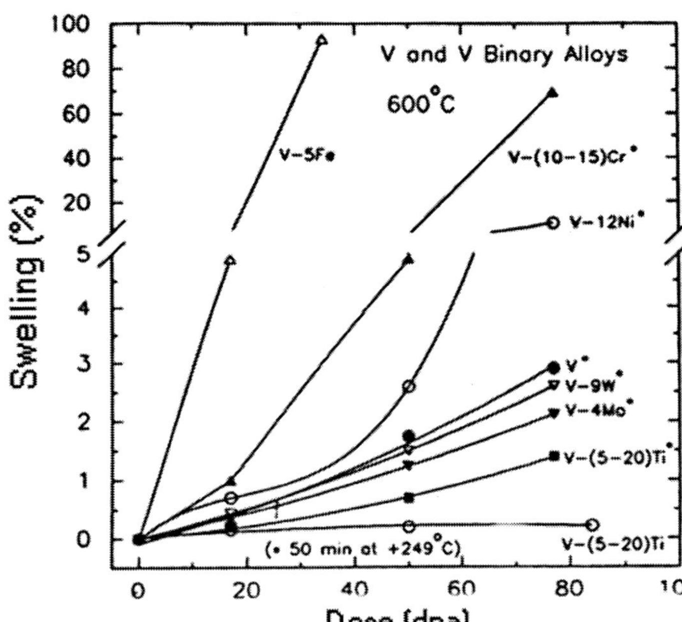

Figure 17. The dose dependences of binary vanadium alloys swelling at 600°C.

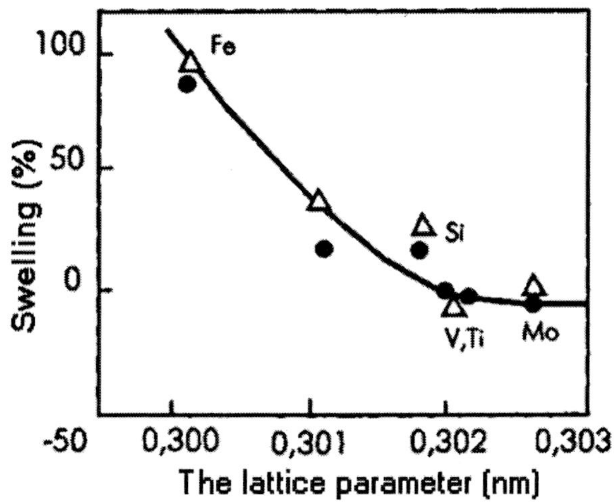

Figure 18. Swelling of vanadium alloys V-5Fe, V-5Cr, V-5Ti, V-5Mo, V-1Si and V, irradiated at 600°C to 45 dpa in the reactor FFTF: •— without helium, 45 dpa; △ — with helium [13].

Figure 19 [35] shows the microstructures of the pores in the alloy V-0.3Fe, irradiated at 520°C and 360/520°C. The bimodal pore size distribution appears in the V-0.3Fe alloy due to the change of irradiation temperature in contrast to the unimodal distribution of pore size at a stable temperature of irradiation. Swelling in both regimes was close to 0.8%. However the major part (~ 90%) of swelling for irradiation of 360/520°C consisted of small pores. The similar microstructural changes associated with the variable exposure, could be seen in V-5Fe, V-4Cr-0.1Ti and V-0.1Fe. The authors note the fact that there is no difference in the evolution of the microstructure of unalloyed vanadium and alloy V-0.3Fe. Nucleation of pores in the alloy V-0.3Fe in the regime 360°C of variable irradiation 360/520°C was essentially more than their growth. The authors also anticipate that shift of the peak of pores nucleation is associated with the differences in the sink power resulting from the additions of undersized solute atoms such as Fe.

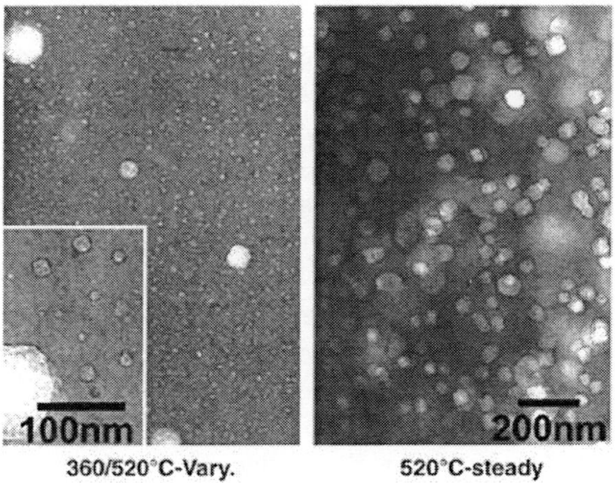

Figure 19. The void TEM microstructures of V-0.3Fe for the 360/520 and 520°C cases [35].

You can once again to note that swelling of the matrix amplified in such solid solutions, in which there are no conditions for the formation of distortions that can serve as sinks of point defects, as it happens, for example, in the event of short-range order in alloys.

Most effectively swelling of vanadium is suppressed by alloying it with titanium within the limits of 5-20%. In investigation [36] using positron annihilation spectroscopy is shown that the dissolved Ti is a very effective trap for vacancies in the alloys of vanadium. Thus, swelling of the alloys V-(5-20) Ti after irradiation at a dose 120 dpa (420°C), 50 dpa (520°C) and 84 dpa (600°C) is less than 0.05%. Chromium supplements (15%) to the V-5Ti alloy increase the swelling, but the magnitude of swelling of alloys V-(0-15)Cr-5Ti after irradiation at a dose 120 dpa (420°C), 50 dpa (520°C) and 84 dpa (600°C) was <0.1%. Swelling of the alloy V-15Ti-10Cr after irradiation in the BN-600 reactor at ~ 450°C and ~ 45 dpa also did not exceed 0.1% [32].

In [37] alloys based on vanadium have been comprehensively evaluated on the basis of their ultimate strength in the unirradiated and neutron irradiated condition, of swelling resistance, the temperature of ductile-brittle transition, the resistance to hydrogen, helium and radiation embrittlement and compatibility with the lithium reactor coolant. On the basis of these evaluations, the alloys V-(0-7) Cr-(5-10) Ti-(0-1) Si are recommended for more intensive assessment to be used as the structural material in the synthesis reactors.

The low swelling of alloys based on vanadium is associated, apparently, with the mechanism of formation during neutron irradiation of short range order and associated lattice distortions in the neighborhood pairs of unlike atoms. In this way is organized nanostructural sublattice of traps of vacancies and interstitial atoms. Analysis of alloys swelling based on of vanadium, as well as for many other transition metals, demonstrate the correlation of the degree of radiation damage reduction with the metastable structures. In the alloys in which do not form nanodomains of short range order (alloys V-5Cr and V-5Fe, V-0,3Fe, V-4Cr-0.1Ti and V-0.1Fe), swelling is very large. At the same time in the alloys of vanadium with titanium and chromium (type V-10Ti-5Cr) can be formed a structure consisting of nanodomains of short range order, and therefore the formation of pores is minimal.

3.4. Compatibility of Vanadium Alloys with Coolants

The solubility of metals in liquid metal coolants (Li, Na, Pb), containing nonmetallic impurities (O, N, C) at low (<0.001%) level increases in the order V, Fe, Cr, Ni. This sequence is maintained for all the considered coolants, and the general level of solubility of materials and respectively the corrosion aggressivity of coolants increases in the order - Li, Na, Pb. The consequence is the fact that in liquid metal coolants of high purity in terms of corrosion resistances decreasing the structural materials are arranged in order: vanadium and its alloys, chromium stainless steels, chromium-nickel austenitic steels. And the level of corrosion losses of these materials in the pure liquid metals increases approximately in the ratio 1: 10: 100. An essential factor determining the corrosion resistance of structural materials in liquid metal coolants is the presence of nonmetallic impurities in the system. For sodium and lead the determining impurity is oxygen. In the sodium of the reactor purity pure vanadium is subjected to the catastrophic dissolution through the formation of easily soluble in sodium compounds Na-O-V. This phenomenon is suppressed by alloying of vanadium with titanium in an amount exceeding 5%. Possessing sufficiently high stability to influence of lead, the vanadium and its alloys are getter with respect to the dissolved oxygen in the liquid metal, which can not expect on the operating temperatures in this system higher than 500-550°C.

Irradiation of vanadium alloys in flowing sodium in a fast reactor BN-600 with neutron fluence of $1.65 \cdot 10^{23}$ n/cm^2 (E > 0.1 MeV) (>45 dpa) at a temperature of 440°C showed [38] that the vanadium alloy specimens decreased substantially in thickness and showed brittle destruction (elongation after irradiation was 0). For alloys with Ti and Cr the specimen thickness did not change, and ductility was retained (Tables 9 and 10).

Table 9. Sample thickness of vanadium alloys before and after irradiation test in BN-600 reactor up to 45 dpa at 440°C in flowing sodium

Alloy composition	Thickness (mm)		Thickness variation (%)
	Before	After	
V-20Ti-10Cr	0.40-0.42	0.40-0.42	0
V-17Mo-0.25Zr-0.16Y	0.36-0.38	015-0.18	50

Table 10. Mechanical properties of vanadium alloys before and after irradiation test in BN-600 reactor up to 45 dpa at 440°C in flowing sodium

Alloy composition	σ_B (MPa)		$\sigma_{0,2}$(MPa)		δ_{total} (%)		δ_{unifom}, (%)	
	Before	After	Before	After	Before	After	Before	After
V-20Ti-10Cr	1070	1080	890	975	9.0	2.5	8.5	2.5
V-17Mo-0.25Zr-0.16Y	770	710	645	-	8.0	0.0	7.5	0.0

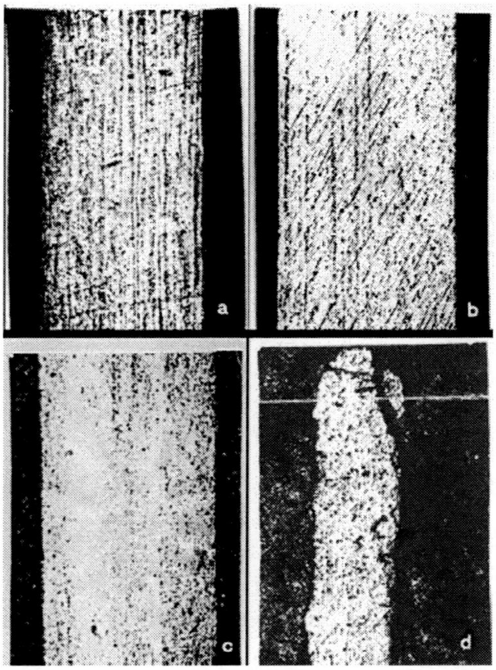

Figure 20. Cross section of samples V-20Ti-10Cr (a, b) and V-17Mo-0, 25Zr-0,16Y (c, d): initial state (a, c); irradiated (b, d) in sodium in the reactor BN-600 at 45 dpa at 440°C, x100.

However, the values of elongation dropped (from 9 to 2.5), which may be explained by the oxynitride film on the specimen surface. Corrosion tests of samples of vanadium alloys in the BN-600 reactor (dose of ~ 45 dpa, the temperature of 440°C) in the sodium coolant flow showed an unsatisfactory resistance of all alloys, except for V-20Ti-10Cr (Figure 20 [1]). However, these results do not guarantee the alloys V-Ti-Cr against degradation at higher temperatures. Chromium stainless steels, especially the ferritic class steel with a chromium content up to 25%, in addition alloyed with silicon and molybdenum, in terms of corrosion resistance are compatible with lead coolant up to 700°C and higher, due to the formation on the surface of structural metal the protective layer of oxides that do not interact with lead.

In sodium of the reactor purity the stainless steels (chromium and chromium-nickel) possess high corrosion stability, and in fact on the corrosion properties the chromium steels exceed the chromium-nickel. For working in the contact with lead at elevated temperatures can be considered only chromium steels. The best corrosion resistance in lead and sodium has the chromium steel such as Fe-17Cr with additives of silicon and molybdenum. The rate of corrosion of this steel is 0.001-0.005 mm / year.

4. The Vanadium Alloys Clad with Ferritic Stainless Steel

To realize the advantages of vanadium alloys on heat resistance and radiation resistance, it is necessary to protect them from the interaction with aggressive media.

For creating coatings on the vanadium alloys as the most promising materials with the best corrosion resistance both in atmospheric conditions and in the liquid-metal coolants environment, as well as in the medium of cooling pond, can be highly corrosion-resistant coatings of chromium ferritic steels with increased chromium content (such as Fe-17Cr, Fe-26Cr). In terms of physical properties : thermal conductivity, the coefficient of linear expansion and, that is especially important, on the parameter of thermal stability vanadium alloys and ferritic steels are similar to each other.

In order to implement the idea of cladding of vanadium alloy has been carried out consistently hot joint (compression) and cold rolling of package from the vanadium alloy and ferritic stainless steel [17]. In this case any bundles or three-layer plate cracking after cold rolling was not observed. Such absence of defects in the metal compounds was the evidence of obtaining strong connection of heterogeneous metals, whereby assembly of ferritic steel - an alloy of vanadium - ferritic steel is deformed as a comprehensive whole.

Subsequent investigations of starting materials connection area (Figure 21) have shown that at temperature 850°C (the temperature of pressing) forms a layer of interaction. The size of the welding zone approximates 6 - 8 pm. It is evident that there is the penetration of vanadium in the steel to a depth of 10 - 12 microns. Annealing the samples of assembly at 1000°C for 2 hours led to the fact that the diffusion layer increased by about 4 - 6 microns and made 10 - 12 microns. In order to determine the quality of connections have been made mechanical testing of three-layer sheet samples at 20, 650, 700 and 800°C in a vacuum of $5 \cdot 10^{-2}$ Pa. One sample was tested at temperature of 700°C in a technological argon environment, which contains up to 6% oxygen and nitrogen impurities. Visual inspection of the tested specimens showed that with the same test conditions (heating rate, the rate of expansion, etc.) the investigated samples are destroyed identically with the formation of the neck. The zone of destruction takes the form oblique cut at an angle of 45° to the longitudinal axis, indicating to a viscous failure (Figure 22). It should be noted that the samples tested in vacuum in the temperature range 20 – 800°C are distinguished with color of a surface.

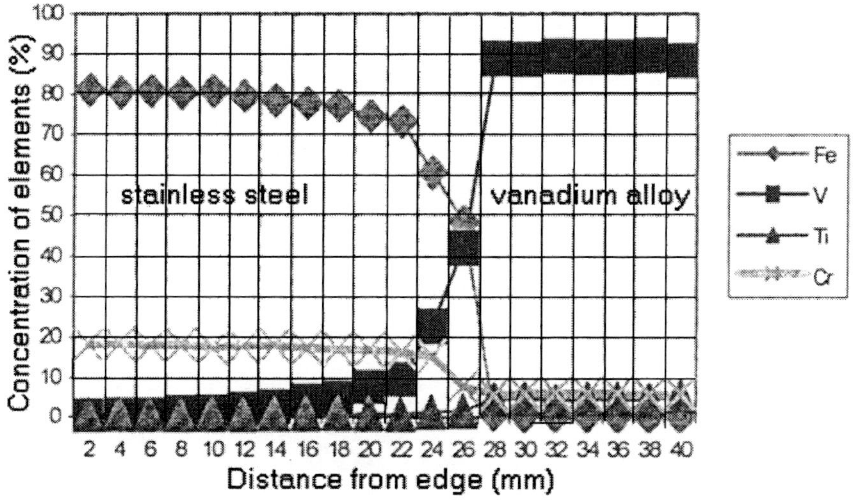

Figure 21. Concentration dependence of the connection area elemental composition after annealing at 850°C for 2 h.

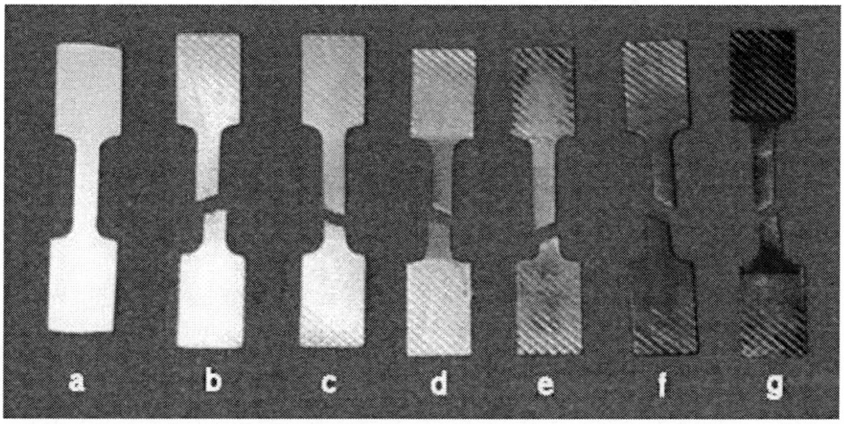

Figure 22. General view of a three-layer sheet before and after mechanical tensile tests : a) Prior to testing, b) 20°C c) 200°C, d) 400°C, e) 650°C, f) 700°C, g) 800°C.

Figure 23 shows the place of the break of specimen after mechanical tensile tests in an argon atmosphere. It can be seen that the sample tested in an argon atmosphere, is also destroyed with the formation of a neck at an angle of 45° to the longitudinal axis. But on the lateral surface of the sample there are tears, that is, there is a slight cracking of protective layer of steel, because the sides of the vanadium alloy was not protected from the environment, which indicates the reaction of vanadium alloy with impurities in argon. When testing under vacuum the cracking was not observed, that confirms the validity of protection of vanadium alloy V-4Ti-4Cr with steel Fe-17Cr. The results of mechanical tensile tests of flat specimens in an argon atmosphere show that the three-layer sheet sample is strengthened in comparison with the specimens tested in vacuum at the same temperature. This indicates that the test conditions strongly influence on the properties of vanadium alloys.

4.1. Obtaining the Bimetallic Tube

For the practical realization of the idea of vanadium alloy tube cladding by corrosion-resistant material has been carried out consistently hot joint (compression) and cold deformation by extrusion of bimetallic assembly. In the assembly for the protection of alloy V-4Ti-4Cr, as it has been tested on flat samples of sandwich plates, used stainless ferritic steel Fe-17Cr. The final size of the obtained bimetallic tube was ∅ 15 x 2.5 mm with a thickness of the cladding layer about 70 microns.

Figure 23. The sample destruction zone after tensile tests in argon at 700°C.

The initial metallographic investigations of the contact zone compound of alloy V-4Ti-4Cr and steel Fe-17Cr in the bimetallic tube have shown that there is the formation of dense welded zone between vanadium alloys and steel. The criterion of the welding quality of alloy and steel work pieces was the lack of pores, bundles, oxides, large nonmetallic inclusions and other defects on the layer boundaries. Therefore we can assume that the quality of welded joint of layers of the bimetallic tube after cold working up to the size ∅ 15 x 2. 5 mm, the fragment of a structure of which is shown in Figure 24, is quite satisfactory. The boundary of zone of the welded joint is clean, smooth, well-marked.

The results of determining the strength properties of the bimetallic tube and its components are shown in Figure 25. Also presented the results of calculations of ultimate strength of the alloy V-4Ti-4Cr in the tube according to the rule of mixture. They show that at room temperature, the value of ultimate strength for the bimetallic tube is higher than for the alloy V-4Ti-4Cr due to the contribution from the layer of steel Fe-17Cr, which under these test conditions, stronger than alloy. This effect persists up to temperatures of about 600°C. At higher temperatures, the bimetal tube has a lower strength than the alloy V-4Ti-4Cr.

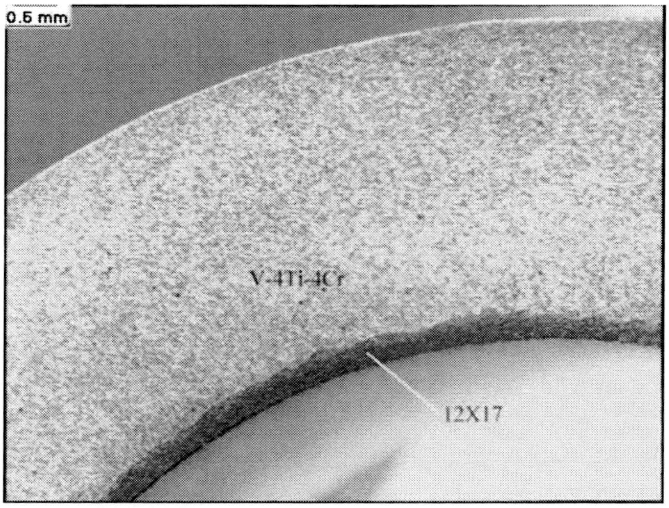

Figure 24. The cross section of bimetallic tube ∅ 15 x 2. 5 mm, alloy V-4Ti-4Cr - steel Fe-17Cr.

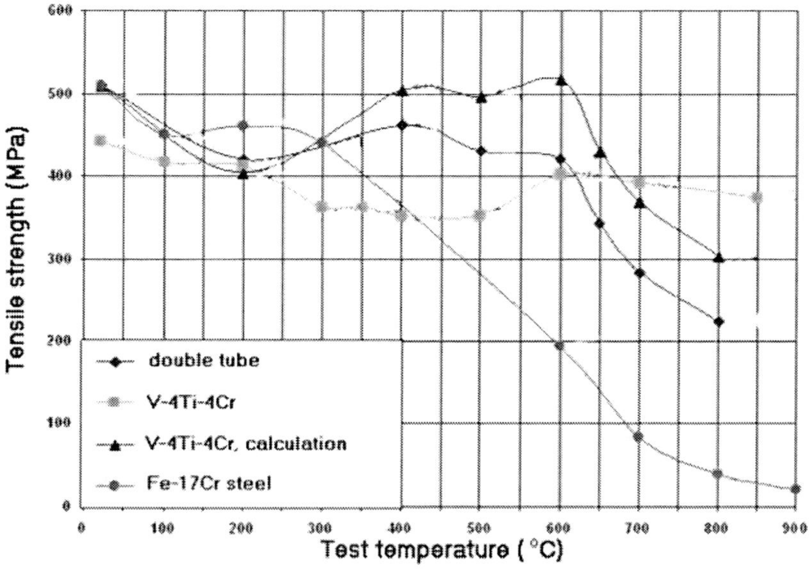

Figure 25. The ultimate strength of the double-tube components vs. temperature.

From the presented brief overview of the properties of structural materials for the fuel rod cladding of active zone of fast reactors it can be concluded that the material that would satisfy all the elevated requirements of prospective new-generation reactors, does not exist in nature. As for the use of vanadium alloys as the fuel rod cladding material of such reactors, where liquid sodium is used as the heat carrier, then the problem solution of using the vanadium-based alloys may be the plating of the shell with ferritic stainless steel [39].

Conclusion

Received a new structural material "Vanfer" for fuel rod claddings of fast reactors, high radiation resistance and the strength properties of the material are provided with V-Ti-Cr alloys, and the corrosion stability - ferritic stainless steels type Fe-17Cr.

References

[1] S.N. Votinov, I.S. Golovnin, V.P. Kolotushkin. Problems of development of advanced materials for the claddings of fast reactors, Proc. Nuclear Power Plants in Russia. *60 years of nuclear industry.* Moscow, 2005, 313-335 in Russian.

[2] K.A. Lanskaya, R.M. Kireeva. *Steels for the boilers of thermal power stations.* "Studies of Heat-Resistant Alloys." Ed. USSR Academy of Sciences, Moscow, 1959. p. 43-50 in Russian.

[3] S.N. Votinov, I.S. Golovnin, V.P. Kolotushkin, S.S. Kislyakov, Y.V. Karasev, A.V. Putilov, O.M. Saraev, V.A. Teplitsky. Vanadium alloy clad ferritic stainless steel - a cladding material of fast neutron reactors. *Proc. Nuclear power stations in Russia.* Moscow, 2007. 317-349 in Russian.

[4] Chang H.M., Loomis B.A., Smith D.L. In *Fusion Reactor Materials:* Semiannual Progress Report: DOE/ER-0313/14, ORNL (1993) 309—317.

[5] Evtikhin V.A., Lyublinski I.E., Vertkov A.V., Votinov S.N., Dedyurin A.I. Influence of alloying and impurity element contents on V-Ti-Cr alloy properties, *J. Nucl. Mat.* 307-311 (2002) 591-595.

[6] D.L. Smith, B.A. Loomis and D.R. Dierks " Vanadium – base alloys for fusion reactor applications, - A review", *J. Nucl. Mat.,* 1985, (135), 125-140.

[7] B. A. Loomis, A. B. Hull and D. L. Smith. Evaluation of low-activation vanadium alloys for use as structural material in fusion reactors, *J. Nucl. Mat.* 179-181, Part 1 (1991) 148-154.

[8] S.N. Votinov, A.I. Dedurin, L.I. Ivanov, I.V. , Altovsky, A.A. Grigoryan, V.A. Evtichin, I.E. Lyublinski, A.V. Vertkov, *Vanadium alloys as structural material of fusion reactors.* VANT, Ser. Fusion, No. 2, Moscow, (2000) 3-12 in Russian.

[9] Altovsky I.V., Borisov A.A., Grigoryan A.A. et al. The project DEMO. *Fundamentals of demonstration fusion reactor DEMO concept.* Part III. Ch. 3. Activation of the DEMO reactor materials. Moscow: NFI RRC KI, (1997) 48-84 in Russian.

[10] S.N. Votinov, L.I. Gomozov, A.I. Dedyurin, L.P. Zavyalsky, Features of vanadium alloys as low-activated materials for the fusion reactors. *VANT, Ser. Fusion,* No. 3, Moscow, (1989) 49-55 in Russian.

[11] S.N. Votinov, V.A. Kazakov, V.A. Evtikhin, I.E. Lyublinski. *IEA Workshop on vanadium alloys for fusion applications*, June 15 – 17, 1994.

[12] Harrot D.L., Gold R.E. Technical properties of vanadium and vanadium – base alloys. – *International metals review,* 4 (1980) 163-221.

[13] S.N. Votinov, A.I. Dedurin, Vanadium – base alloys - a promising low activation structural material for fusion reactors. *VANT. Ser. Material science and new materials,* No. 1 (52) (1995) 66-73 in Russian.

[14] Shunk F. *Constitution of binary alloys* .- Moscow: Metallurgiya, 1973. 760 in Russian.
[15] Fromm E., Gebhard E. *Gases and carbon in metals* / translated from German .- Moscow: Mir, 1980. 493.
[16] Efimov Y.V., Baron V.V., Savitsky E.M. *Vanadium and its alloys.* Nauka, Moscow, 1969, 1—254 in Russian.
[17] I.E. Lyublinski, A.V. Vatulin, S.N. Votinov, I.N. Gubkin, Y.V. Karasev, G.P. Vedernikov, V.A. Evtihin, A.V. Vertkov, A.I. Dedyurin, I.V. Borovitskaya. Optimization of the alloying of V-Ti-Cr alloys, *VANT, Ser. Fusion,* No 3 (2005) 70-78 in Russian.
[18] O.A. Alekseev, S.N. Votinov, I.N. Gubkin, Yu.V. Karasev, V.P. Kolotushkin P., S.A. Nikulin, L.V. Potanina, S.G. Sergeev, D.V. Sokolovsky, Vanadium alloy clad ferritic stainless steel - the fuel rod cladding material of fast reactors, *Advanced Materials,* № 4 (2009) 34-42 in Russian.
[19] S.N. Votinov, V.P. Kolotushkin, O.V. Skiba. Vanadium base alloys as a fuel rod cladding material for fast breeder reactors, *Proc. Nuclear Power Plants.* 20 years after the Chernobyl accident. Moscow, 2006, 303-316 in Russian.
[20] Votinov S.N., Dedurin A.I., Kazakov V.A. et al. Development of vanadium alloys for fusion blankets. *Proceeding of the IEA Workshop on vanadium alloys for fusion application.* Salem, Oregon, USA, June 15-17 1994.
[21] H. Watanabe, T. Muroga and N. Yoshida. The study of temperature variation during HFIR irradiation on vanadium, *J. Nucl. Mater.* 329-333, Part 1 (2004) 425-428.
[22] Smith D.L., Chang H.M., Loomis B.A., Matsui H., Votinov S.N., Van Witzenburg W. Development of Vanadium – Base Alloys for Fusion First Wall/Blanket Applications. The Third International Symposium of Fusion Nuclear Technology, June 27- Jule 1, 1994, UCIA Conference Center, - Los-Angeles, California.
[23] B.A Loomis, D.L. Smith. Effects of neutron irradiation, helium and hydrogen on ductile-brittle transition temperatures in vanadium alloys. *Sixth International Conference on Fusion Reactor Materials,* Stresa, Lago Maggiore. Italy. 27 September, 1 October, 1993.
[24] S.N. Votinov. V.A. Evtikhin. Lyublinski. A.Y. Yertkov. Selection of vanadium alloy for high-temperature liquid lithium blanket of fusion reactor. *Proceeding of IEA/ JUPITER-II Joint International Workshop on Liquid Blanket and Low Activation Material System.* 21-23 Mai 2001, Sendai. Japan.
[25] *The phase diagrams of binary metallic systems.* Volume 3. Book 2. Moscow: Machinery, (2001) 397-399 in Russian.
[26] Molokanov V.V., Budberg P.B. *Phase structure of the Ti - V alloys system, Reports of the Academy of Sciences,* 215 No. 5 (1974) 1125-1127 in Russian.
[27] V. P. Kolotushkin, V. N. Rechitskii, O. B. Ermolova, and A. V. Laushkin, "Effect of a Structure–Phase State on the Plastic Properties of Nickel–Chromium– Molybdenum Alloys," *VANT. Ser. Material science and new materials,* 1(61) (2003) 94–104 in Russian.
[28] J.M. Chen, T. Muroga, T. Nagasaka, Y. Xu, C Li, S.Y. Qiu, Y. Chen. Precipitation behavior in V-6W-4Ti, V-4Ti and V-4Cr-4Ti alloys //J. Nucl. Mater. 334 (2004) 159–165.
[29] S.S. Batsanov. *Structural Chemistry. Facts and dependences.* Moscow. 2000. 1-291. P. 18.

[30] *The phase diagrams of binary metallic systems*. Volume 2. Moscow: Machinery, (1997) 193-195 in Russian.

[31] J.M. Chen, T. Muroga, T. Nagasaka, S.Y. Qiu, C Li, Y. Chen, B. Liang, Z. Y. Xu. The mechanical properties of V–4Cr–4Ti in various thermo-mechanical states, *Fus. Eng. and Des.* 81. Issues 23-24 (2006) 2899–2905.

[32] Dedyurin A.I., Borovitskaya I.V., Dvoryashin A.M. Effect of irradiation in a fast reactor BN-600 on the structure and mechanical properties of vanadium alloys. *Proceedings of the IX International Meeting,* "Radiation Physics of Solids" Sevastopol, June 28 - July 3, 1999, Vol 1, Moscow (1999) 1212-1225 in Russian.

[33] T. Hayashi K. Fukumoto and H. Matsui. Effect of undersized solute atoms on point defect behavior in V–A (A=Fe, Cr and Si) binary alloys studied by using HVEM, *J. Nucl. Mater.*. 307-311, Part 2 (2002) 930-934.

[34] H. Matsui, and D.S. Gelles. *Large swelling in V-5Fe alloy after irradiation in FFTF.* ANL-1989, p. 112.

[35] K. Fukumoto, H. Matsui, T. Muroga, S.J. Zinkle, D.T. Hoelzer, L.L. Snead. Varying temperature effects on mechanical properties of vanadium alloys during neutron irradiation, *J. Nucl. Mater. 329-333 (2004) 472–476.*

[36] T. Leguey, R. Pareja and E.R. Hodgson. Annealing of radiation-induced defects in vanadium and vanadium-titanium alloys, *J. Nucl. Mater.* 231, Issue 3 (1996) 191-198.

[37] B. A. Loomis, A. B. Hull and D. L. Smith. Evaluation of low-activation vanadium alloys for use as structural material in fusion reactors, *J. Nucl. Mater.* 179-181, Part 1 (1991) 148-154.

[38] S.N Votinov et al., Influence of irradiation in fast BN-600 reactor on structure and mechanical properties of vanadium alloy, in: *Proceeding of Ninth International Symposium on Solid State Radiation Physics.* Sevastopol. 28 June- 3 July 1999. Moscow, vol. 2. 1999. p. 1212 (in Russian).

[39] The patent for the invention № 2331941 *The shell of fuel element fast reactor with liquid metal coolant.* The application number 2006135586 Priority of invention October 9, 2006 Registered in the State Register of Inventions of the Russian Federation August 20, 2008 The patent term expires on October 9, 2026.

In: Vanadium
Editors: V. N. Baranova and A. V. Fortunatov
ISBN 978-1-62081-600-4

Chapter 4

VANADIUM OXIDES FOR SOLAR CONTROL APPLICATIONS

Dimitra Vernardou[1] and Martyn E. Pemble[2]
[1]Center of Materials Technology and Photonics,
School of Applied Technology,
Technological Educational Institute of Crete, Crete, Greece
Science Department, School of Applied Technology,
Technological Educational Institute of Crete, Crete, Greece
Department of Materials Science and Technology,
University of Crete, Crete, Greece
[2]Tyndall National Institute, University College Cork,
Lee Maltings, Prospect Row, Cork, Ireland

ABSTRACT

Solar control window coatings are those which can respond in a desired way to an external stimulus. These are materials such as the vanadium dioxide that can control the amount of daylight and solar heat that can pass through the windows of buildings with temperature.

Chemical vapour deposition is an attractive method for thin film deposition since it is capable of achieving high deposition rates. The properties of the as-deposited samples can be simply tuned by altering the vapour composition and flow in the coating zone. This can be beneficial for the thermochromic coatings since stoichiometric vanadium dioxide coatings are required. Finally, when CVD is performed at atmospheric pressure, it has the advantage of a relatively simple apparatus (no vacuum system required), which can be easily integrated into float-glass production lines.

Results obtained demonstrate the excellent thermochromic properties of undoped vanadium dioxide coatings as illustrated by switching temperature as low as 51 °C. Finally, from the in-situ monitoring analysis it can be possibly concluded that in the absence of oxygen, anhydride species may form with some evidence of ester species. While, in the presence of oxygen, there is indication for the enhanced production of CO and the suppression of the C-H species.

1. Background

The use of air conditioning systems to maintain comfortable working environments in modern times has become increasingly common. This leads to a dramatic increase in the use of electricity and the concurrent increase in CO_2 emissions and other atmospheric pollutants that are released in the electricity generating process. Solar control coatings such as thermochromics on windows is a technology being developed that if widely employed could enable a reduction in the use of air conditioning systems, and hence a reduction in energy costs and greenhouse emissions.

Thermochromic coatings are thin films coated on the glass panes of a window, which heat reflectivity changes with the temperature of the glass. Thus, these windows could have a significant benefit to the consumer. For example, above T_c (transition temperature), they are more IR reflective, preventing thermal radiation from excessively heating the building interior, thus ensuring that the inside of the building does not overheat during the summer (figure 1.1). It has been known since the late 1950's that VO_2 displays thermochromic behavior [2]. It undergoes a fully reversible metal-semiconductor phase transition (MST), which is associated with drastic changes in electrical and optical properties. Above T_c (transition temperature), VO_2 has a tetragonal rutile type structure (metallic phase) characterized by chains of edge sharing [VO_6] octahedral along the c-axis with equidistant vanadium atoms (V-V = 2.88 Å) [3]. While, below T_c, a monoclinic structure (semiconducting phase) is presented by pairs of vanadium atoms V^{4+}-V^{4+} with alternate shorter (0.265 nm) and longer (0.312 nm) distances, which move together to form diatomic molecules [4]. The monoclinic and tetragonal structures, where oxygen and vanadium are presented by red and white spheres respectively, are indicated in figure 1.2.

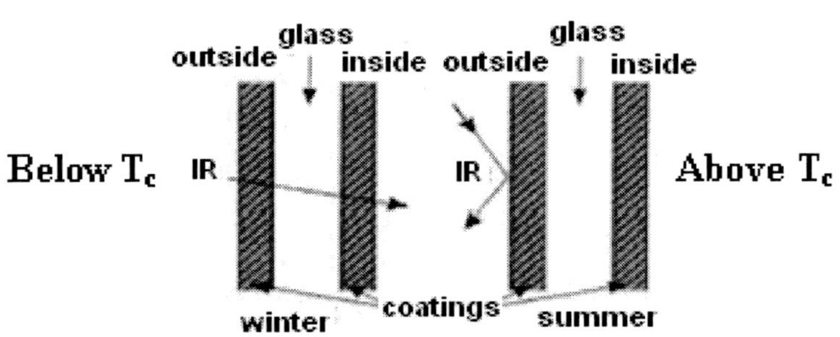

Figure 1.1. A schematic presentation of the thermochromic windows [1].

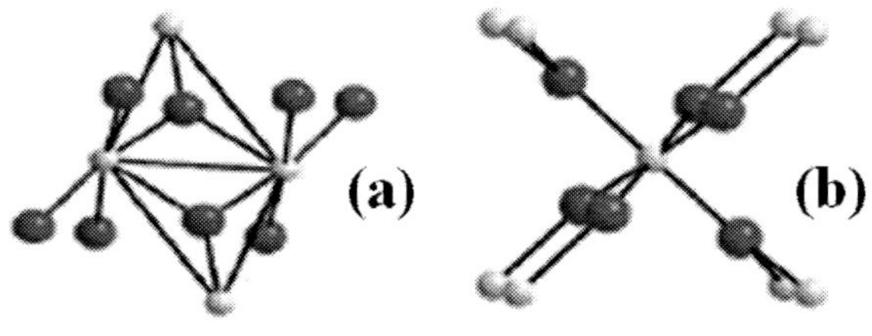

Figure 1.2. Monoclinic (a) and tetragonal (b) VO_2 [5].

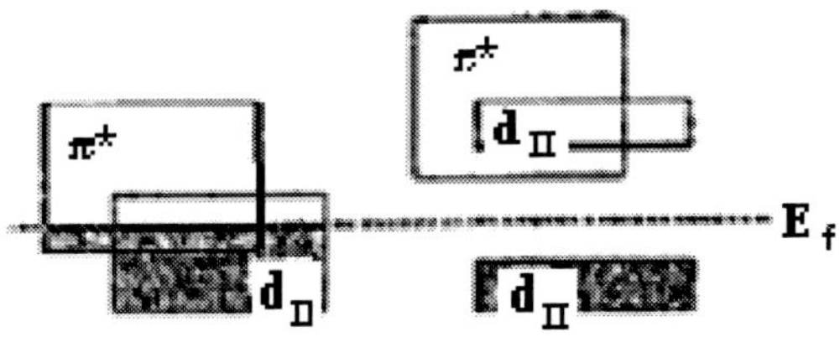

Figure 1.3. Schematic sketch of the band structure of single crystalline VO_2 in the metallic and semiconducting phases after Goodenough (where E_F is the energy at Fermi level).

Goodenough [6] has proposed that the thermochromic effect can be attributed to the changes in bonding with the phase transition. Figure 1.3 shows the π^* and $d_{||}$ bands resulting from the t_{2g} levels of the oxygen octahedral overlap at the Fermi levels, thus being responsible for the metallic properties above T_c. In the semiconducting phase V-V pairing and off-axis displacement occurs along the vanadium chains resulting in a monoclinic structure. The V-O hybridization is changed, the V-V bonding is stronger and the π^* band rises above the Fermi level and the $d_{||}$ band is split into an empty and a filled orbital as indicated in figure 1.3.

In order for VO_2 to find use in high-volume applications, it is desirable that the T_c should be reduced (< 68 °C) to be applicable in different global locations. In this respect, factors such as crystalline quality, thickness [7] and doping depending on parameters such as size [8], charge of the dopant ion [9], and changes in electron-carrier density [10] may significantly reduce T_c.

Chemical vapour deposition (CVD) routes are more attractive than other possible methods for the production of VO_2 coatings, since the stoichiometry of the films may be precisely controlled through the growth parameters such as substrate temperature and oxygen flow rate. In addition, the simplicity of CVD when performed at atmospheric pressure (APCVD) would make such a process compatible with high-volume, in-line glass manufacturing processes. As a result, a number of studies of the growth of VO_2 thin films have been reported. The first of these concerns the use of vanadium (IV) chloride (VCl_4) with methanol or ethanol to prepare V_2O_5, which was subsequently post-reduced to VO_2 by heating in a reducing atmosphere. However, the films were significantly contaminated with carbon [11]. Vanadium (V) oxytrichloride ($VOCl_3$) has also been used along with water to produce VO_2 films. In this case, high temperatures (ca. 600 °C) were found to be necessary with water present in excess in the gas phase [12]. Furthermore, it has been shown that VO_2 can be prepared directly or by post- reduction of V_2O_5 using vanadyl tri(isopropoxide) [13] or vanadyl tri(isobutoxide) [14] as single-source precursors. Regarding the vanadyl compound $VO(acac)_2$ has been used to prepare the metastable VO_2(B), which was then converted to the tetragonal phase VO_2 by annealing in argon at 500 °C [7,15].

A comparison of the literature available reveals that various conditions have been used in terms of reactor types, flow rates, temperatures studied and precursor concentrations. However, it has not been a validate comparison among the methods apart from one case [16] and a systematic study to state what the critical growth parameters are in each method. Furthermore, the mechanisms of gas phase precursor reactions and reactive intermediate production are poorly understood in all cases.

The present chapter addresses this point directly, using a "standard" APCVD reactor system of the type used in the glass industry to mimic the conditions likely to be found in a commercial glass-coating environment. We present data that clearly identify the useful CVD parameter space windows where VO_2 films possessing the desired qualities. The critical importance of achieving the optimum morphology and oxidation state of the vanadium oxide is highlighted. We finally comment on the gas phase reaction mechanisms taking place in the APCVD system using in-situ Fourier transform infrared spectroscopy studies.

2. EXPERIMENTAL

The vanadium oxide thin films were grown using an APCVD reactor of in-house construction as shown in figure 2.1 [16]. Figure 2.1 indicates the reactor in which the carbon susceptor is heated by three heating pencils to temperatures up to 600 °C. On top of this lies the substrate that is to be coated, which sits approximately 10 mm below a glass ceramic top plate. On the right of the reactor is the exhaust where all of the waste gases are carried and taken directly to waste through the extraction system. One can also see the baffle, which causes the gas (carrier gas nitrogen mixed with precursor) to flow in a laminar manner across the substrate. The APCVD rig was designed with all the outlet lines having no joints to avoid blocking as well as a four-way valve was preferred instead of a three-way valve for easiness in operation. All heaters were regulated by Eurotherm controllers and RS type k thermocouples. A series of films was produced using $VO(acac)_2$ as precursor, which was placed in a bubbler of the APCVD rig. A flow of nitrogen gas was used as the process gas in all reactions, while high purity oxygen gas was used as the oxidation source during the reactions. The oxygen flow rate was 0, 0.2, 0.4, 0.6 and 0.8 L min^{-1} over a substrate temperature range of 450 – 575 °C. The bubbler was heated at 200 °C, while the gas lines were heated at 220 °C to enhance transport of precursor and to prevent condensation in the lines. The four-way valve was kept at 200 °C. The deposition time was 10 min and the nitrogen flow rate through the bubbler (with the $VO(acac)_2$) was 1.4 L min^{-1} for total gas-flow rate of 12 L min^{-1}. Finally, the substrate used was commercial SiO_2-precoated glass of dimensions 220 mm x 85 mm x 3 mm. All substrates were cleaned with H_2O and detergent, rinsed thoroughly with H_2O and deionised H_2O, then allowed to dry prior to deposition.

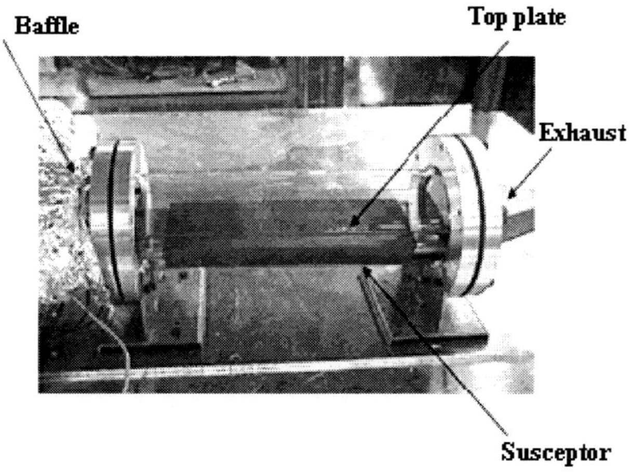

Figure 2.1. Photograph of the cold-wall in-house APCVD reactor.

X-ray diffraction (XRD) measurements were performed using a Siemens D5000 Diffractometer under operating conditions of 40 kV, 40 mA for 2θ = 10.000-50.000, step size 0.1 ° and time step 30 sec / o. The measurements at different temperatures were obtained under the same conditions using a designated heating tape. Temperature-dependent Raman measurements were carried out using a Renishaw 1000 spectrometer with a 632.8 nm laser operating at 2 mW power, 50 % magnification, wavenumber range of 100-1000 cm^{-1} and 50 accumulations. In this case, the sample temperature was controlled using a Linkam THMS600 variable temperature cell. Rutherford backscattered spectra (RBS) were obtained using a 2 MeV accelerator, the analyzing beam was 2 MeV He^+ with the sample being analysed at normal incidence with a scattering angle of 168 ° in IBM geometry. Quark software was used for the simulation.

Regarding the x-ray photoelectron spectroscopy (XPS), it was performed on a XSAM800 instrument using unmonochromated MgKa radiation as X-ray source with line width of 0.8 eV. An argon-sputtering gun used to clean the surfaces of carbon for 150 sec at 3 KeV. The E_B values were referenced to C1s peak at 285 eV. XPS data was analysed using CasaXPS software version 2.2.29. Scanning electron microscopy (SEM) was performed on a Jeol JEM-100CX electron microscope with a tilt angle of 80 °. Samples were over-coated with a thin film of gold to make them more conductive prior to measurements. In terms of the spectral transmittance in normal incidence and the spectral reflectance in 30 ° incidence angle and S polarization in the IR region as well as the visible region were measured at 25 °C and 65 °C using a designated heating tape. In addition, the reversible changes in optical transmittance versus temperature were performed on a Perkin-Elmer 457 grating spectrometer set to 2.5 μm.

An aluminium temperature cell controlled by RS heaters, Eurotherm temperature controllers and k-type thermocouples were used to control the sample temperature. Finally, in-situ monitoring measurements were performed using an Equinox 55 spectrometer under operating conditions of: mid-IR region 500-4000 cm^{-1}, resolution 1 cm^{-1}, sample scan time and background scan time of 800 scans. The beam splitter was KBr.

A schematic presentation of the optical arrangement for the IR measurements is shown in figure 2.2 [17]. Figure 2.2 shows the spectrometer where the IR beam falls on Mirror 1, which is on the level of the machine. The beam is then directed upwards to Mirror 2, which is on the level of the reactor. This directs the beam through the window 1 and 2 of the reactor tube, and then onto Mirror 3, which focuses the beam onto the detector.

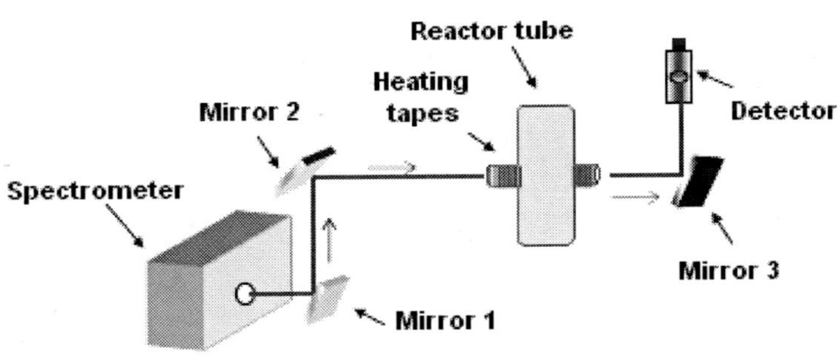

Figure 2.2. Detailed presentation of the optical arrangement used for the FTIR monitoring experiments. The arrows indicate the direction of the beam.

3. Film Characterization

The vanadium oxide films produced had the same thickness profiles over the whole substrate. The thickness did not appear to be affected by the variations in temperature across the substrate. For a given sample, the phase of the vanadium oxide was found to be the same over the entire substrate, i.e. variations in phase as a function of position on the substrate were not observed.

The oxidation phase of the films grown on commercial SiO_2-precoated glass for various oxygen flow rates and substrate temperatures are shown in phase / parameter diagram presented in figure 3.1. The data shown in figure 3.1 was obtained from XRD and Raman spectroscopy. It was found that there was no film deposition at substrate temperatures below 500 °C and even at higher temperatures, addition of oxygen was necessary to induce growth of the vanadium oxide films.

Quantitative surface and near surface analysis was carried out on all samples using RBS. Figure 3.2 shows typical measured (orange circles) and computer-simulated (red line) data for both the substrate and the vanadium dioxide film. This figure presents that the back edge of the vanadium signal is sharp, indicative of a sharp interface between substrate and film. There was no carbon contamination in the bulk of the vanadium oxide films grown at any oxygen flow rates and substrate temperatures studied.

Surface composition of vanadium oxide films has been estimated using XPS data. There was no carbon contamination detected on the surface of the films for any oxygen flow rate over the range of substrate temperatures employed. Figure 3.3 shows the deconvoluted spectra of the $V2p_{3/2}$, $V2p_{1/2}$ and O1s peaks. The $V2p_{3/2}$ peak is generally found to be more sensitive to phase change, and it was used to identify the phase of vanadium oxides [18]. These data indicate that the vanadium in this film is in a single oxidation state of V^{4+} at E_B of 515.9 eV with FWHM of 3.8 eV.

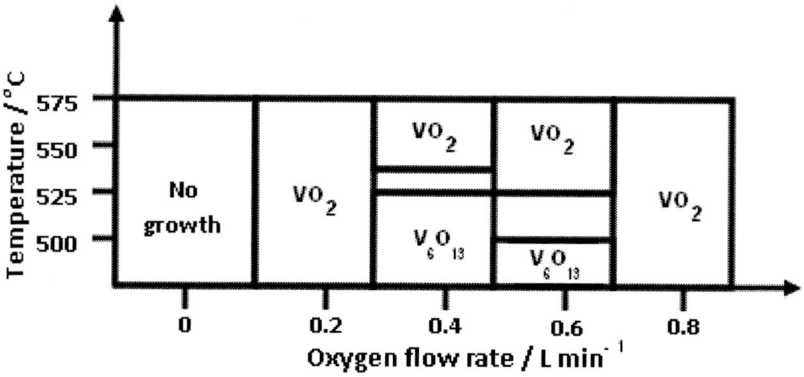

Figure 3.1. Phase / parameter diagram showing the influence of substrate temperature and oxygen flow rate on the oxiadation phase of the films deposited on commercial SiO_2-precoated glass (obtained from XRD and Raman spectroscopy) in an APCVD reactor using $VO(acac)_2$ at deposition time of 10 minutes and a total gas-flow rate of 12 L min^{-1}.

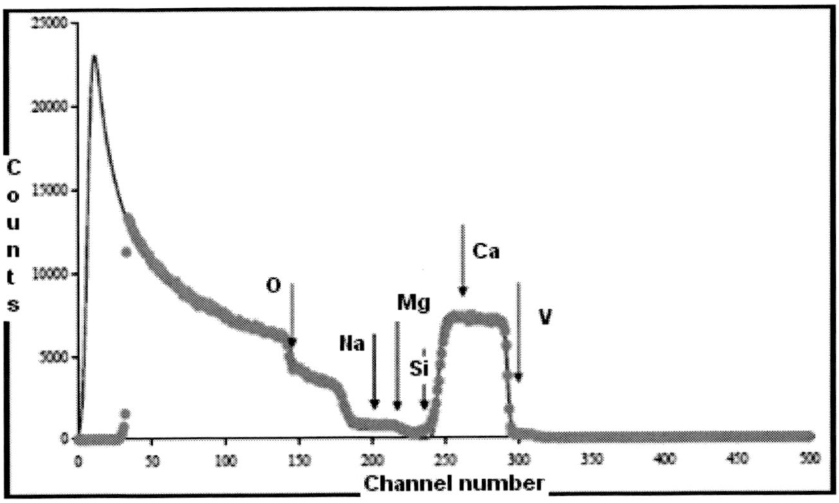

Figure 3.2. RBS data for VO_2 film deposited in an APCVD reactor, showing both the experimental (orange circles) and simulated data (red line). Growth was performed at 525 °C for oxygen flow rate of 0.8 L min^{-1}, deposition time of 10 minutes and a total gas-flow rate of 12 L min^{-1}.

Regarding the $V2p_{1/2}$ peak has been assigned at 522.7 eV with FWHM of 6.2 eV. The main O1s line at 529.9 eV with FWHM of 2.1 eV can be attributed to O^{2-} ions. These E_B values are in good agreement with the values obtained by other researchers [19,20]. There is also a less intense peak at 532.2 eV with FWHM of 1.5 eV, which is assigned to be either due to an amount of carbon bonded with oxygen on the surface of the film [21] or due to the presence of chemisorbed OH groups over the film [22]. Since there is no other carbon indication on the spectra, is then more likely to be due to chemisorbed OH groups.

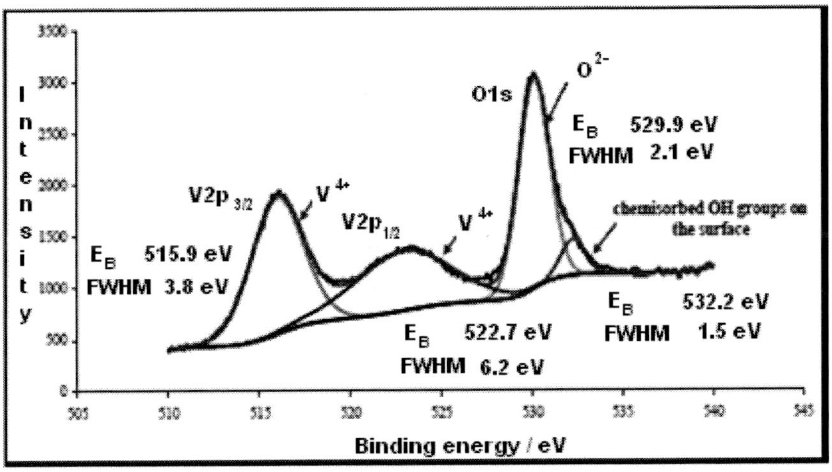

Figure 3.3. XPS deconvoluted spectra of VO_2 films deposited in an APCVD reactor using oxygen flow rate of 0.2 L min^{-1} at 525 °C for growth time of 10 min and a total gas-flow rate of 12 L min^{-1}. The red line represents the raw data, while the black line represents the deconvoluted fit.

Figures 3.4 (a), (b) and (c) show SEM images of vanadium oxide samples grown at 500, 525 and 575 °C respectively. A rough surface, consisting of large, closely-compact and well-formed crystallites of a nearly rectangular shape is observed for vanadium dioxide deposited at 575 °C, while the growth rate is estimated to be ca. 31 ± 10 nm min^{-1}. Decreasing the growth temperature to 525 °C, the crystallite average grain size reduces, while the growth rate is found to be ca. 18 ± 10 nm min^{-1}. Further decreasing the substrate temperature to 500 °C results in a further reduction in the crystallite average grain size and the appearance of voids in between the grains, while the growth rate is calculated to be ca. 15 ± 10 nm min^{-1}.

As expected for a conventional activated process, increasing substrate temperature results in an increase in growth rate. Thus, the films appear to be more crystalline, free of voids and with well-defined grains.

Figure 3.4. SEM images of vanadium oxide films grown in an APCVD reactor for growth time of 10 min, total gas-flow rate of 12 L min^{-1} and oxygen flow rate of 0.2 L min^{-1} at 500 (a), 525 (b) and 575 °C (c). The line is equal to 0.4 micron. The tilt angle is 80°.

SEM images of vanadium oxide samples grown using oxygen flow rates of 0, 0.4 and 0.8 L min^{-1} at 550 °C are shown in figures 3.5 (a), (b) and (c) respectively. Figure 3.5 (a) shows an extremely flat, featureless surface that appears to be free of grain boundaries. The growth rate is estimated to be 68 ± 12 nm min^{-1}. As the oxygen flow rate increases to 0.4 L min^{-1}, a granular surface composed of closely packed and dome-shaped crystal grains of various sizes is presented with a growth rate of ca. 38 ± 12 nm min^{-1}. Regarding the 0.8 L min^{-1} sample, similar morphology to that of 0.4 L min^{-1} is shown, but with smaller and irregular grains with a growth rate of ca. 25 ± 12 nm min^{-1}.

From these data, it is clear that there is a strong dependence of the growth rate on oxygen flow rate suggesting that there is some type of pre-reaction taking place between the precursor and the oxygen.

Figure 3.5. SEM images of vanadium oxide films grown in an APCVD reactor for growth time of 10 min, total gas-flow rate of 12 L min^{-1} and oxygen flow rate of 0 (a), 0.4 (b) and 0.8 L min^{-1} (c) at 550 °C. The line is equal to 1 micron. The tilt angle is 80 °.

4. THERMOCHROMIC BEHAVIOUR

The thermochromic properties and switching characteristics of the films have been evaluated using the following techniques:

1. XRD data were obtained above and below the T_c.
2. Raman spectra were recorded at different temperatures in order to specify the switching characteristics of the films.
3. Transmission and reflectivity of the films were obtained above and below T_c, for the UV-visible and the IR-region.
4. The change of transmittance was recorded over a range of temperature at 2.5 μm in order to determine the T_c value of the films and examine the reversibility of the transition.

Each of the characterization methods was performed on at least three samples prepared under the same conditions for consistency and reproducibility. The phase transition was also multiply cycled up to three times without any apparent variation in T_c or film degradation for the transmittance-temperature and Raman spectroscopy measurements.

The thermochromic transition of VO_2 was observed from the monoclinic-to-tetragonal structural change in the XRD pattern for a range of temperatures, 25, 35, 45 and 65 °C. Above the T_c, the conversion of the VO_2 monoclinic phase to the VO_2 tetragonal phase was observed as shown in figure 4.1. The transition resulted in the removal of the (011) reflection from the monoclinic phase and the appearance of the (110) reflection due to tetragonal phase at lower 2θ. Similar observations have been obtained by other researchers [23,24]. The same pattern was observed for all VO_2 films deposited under different substrate temperature with increased peak intensity for higher oxygen flow rate employed during growth, indicative of a gradual improvement in crystallinity.

The phase transition was also studied using Raman spectroscopy. A set of spectra is shown in figure 4.2 where a VO_2 film is seen to possess a T_c of below 65 °C. As the temperature of the film is increased, the Raman peaks of the monoclinic phase decrease significantly. While as the film is cooled, the Raman bands of monoclinic VO_2 reappear illustrating the reversible nature of the transition. Other researchers have observed similar effect of temperature on band intensity, however the mechanism for such effect remains unclear [25].

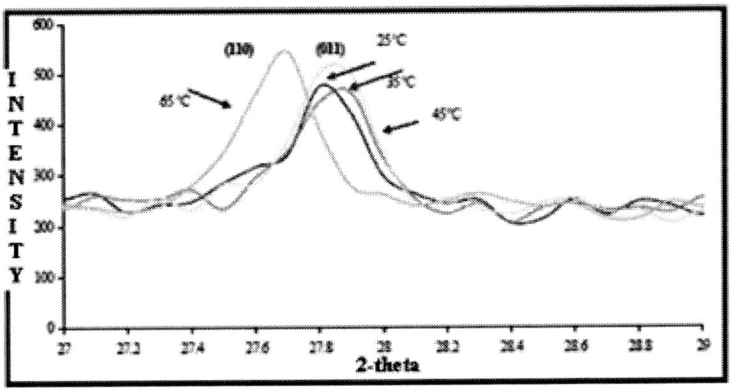

Figure 4.1. XRD patterns as a function of temperature over the range of 25 °C to 65 °C showing the transition between the monoclinic (011) reflection to the tetragonal (110) reflection, for a VO_2 thin film grown at 525 °C.

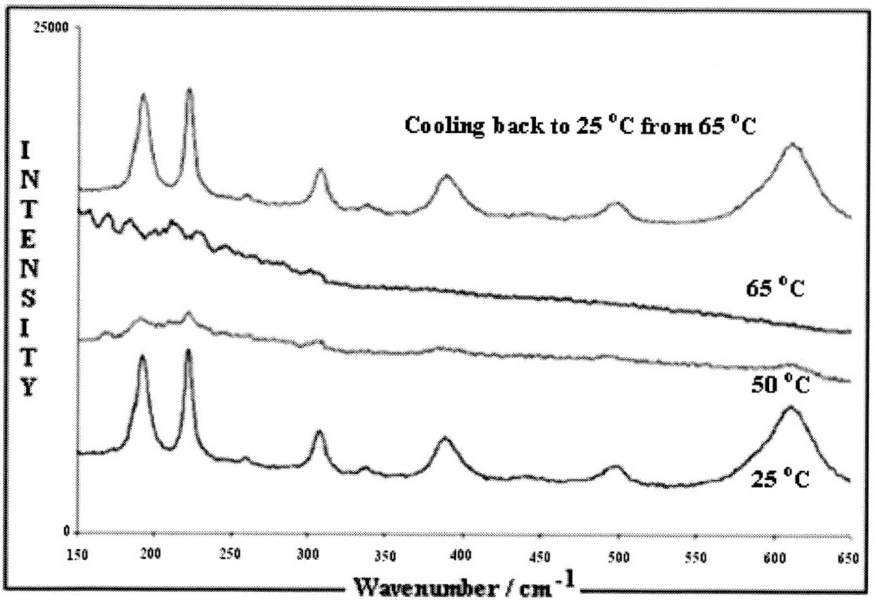

Figure 4.2. Raman spectra recorded at 25, 50 and 65 °C and cooling back to 25 from 65 °C. The film was grown using oxygen flow rate of 0.8 L min^{-1} at 525 °C.

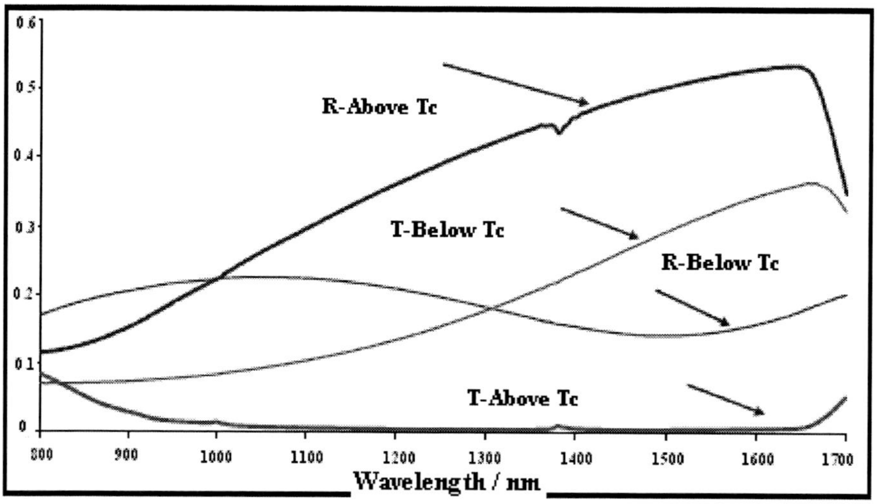

Figure 4.3. Transmittance and reflectance profiles for a VO_2 film at 25 °C (below T_c) and 65 °C (above T_c) over the near to mid-IR region of 800-1700 nm. The film was grown using oxygen flow rate of 0.6 L min^{-1} at 525 °C.

Figure 4.3 shows the transmittance and reflectance of the vanadium dioxide sample over the near to mid-IR region (800-1700 nm), recorded at 25 °C (below T_c) and 65 °C (above T_c). The film displays the basic properties required for intelligent window applications, in the form of general IR transmittance and reflectance characteristics, i.e. increased reflectance (reduced transmittance) in the spectral region 800-1700 nm with no or only slight changes in the visible region 350-700 nm.

The n and k over the IR spectral region (800-1700 nm) of the VO_2 film were determined from the measured transmittance, reflectance and the film thickness (obtained from SEM) in both the semiconducting at 25 °C (below Tc) and the metallic phase at 65 °C (above T_c) as shown in figures 4.4 (a) and (b) respectively. The theoretical model used to describe the variation of n and k was the Gauchy model [26]. This determination employs an approach in which the system is considered as a film deposited onto an optically smooth, transparent substrate [27]. Although, the VO_2 films appeared rough in SEM, they were sufficiently uniform to permit the reproducible determination of optical constants. From figures 4.4 (a) and (b), it is seen that n decreases and k increases when heating the sample from 25 °C (semiconducting phase) to 65 °C (metallic phase), which demonstrates the applicability of VO_2 thin films for use as window coatings.

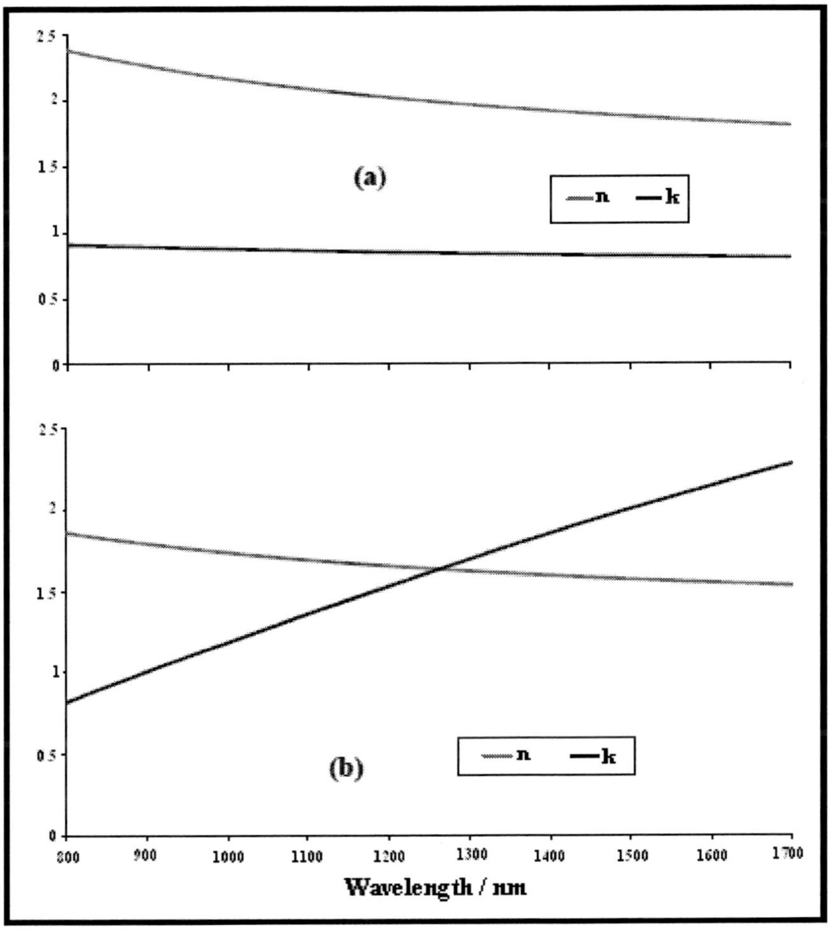

Figure 4.4. Calculated n and k values of a VO_2 film in the semiconducting phase (25 °C) (a) and the metallic phase (65 °C) (b) over the IR spectral region of 800-1700 nm. The film was grown using oxygen flow rate of 0.6 L min^{-1} at 525 °C.

Table 4.1 illustrates the thermally induced reversible phase transition phenomenon during heating and cooling of vanadium oxide thin films, which were prepared at different oxygen flow rates and substrate temperatures. The thermochromic effect may be quantified by the difference in transmittance and reflectance (at a wavelength of 1500 nm) measured at 25 °C and 65 °C. The largest percentage changes in transmittance and reflectance are for the film grown at 525 °C using an oxygen flow rate of 0.6 L min^{-1}, being 28.9 % and 36 % respectively.

Table 4.1. Transmittance and reflectance values of vanadium oxide films deposited in an APCVD reactor recorded at 1500 nm, at 25 °C (below T_c) and 65 °C (above T_c). Films were grown using oxygen flow rates of 0.2 L min^{-1} and 0.6 L min^{-1} at 500 °C, 525 °C and 550°C

	Oxygen Flow Rate / L min^{-1}							
	0.2				0.6			
	T		R		T		R	
	25 °C	65 °C	25 °C	65 °C	25 °C	65 °C	25 °C	65 °C
500 °C	0.291	0.015	0.503	0.605	0.381	0.035	0.332	0.383
525 °C	0.282	0.022	0.402	0.558	0.294	0.005	0.144	0.504
550 °C	0.275	0.015	0.257	0.479	0.353	0.018	0.265	0.356

It is seen that the poorest thermochromic effect is observed for samples grown using an oxygen flow rate of 0.6 L min^{-1} at 500 °C with percentage changes in transmittance and reflectance being 34.5 % and 5 % respectively. It is suggested from the data presented in the phase / parameter diagram (figure 3.1) that the reason for the poor thermochromic properties is most likely that some oxygen rich phase was also formed (e.g. a phase similar to V_6O_{13}).

It is observed that the switching characteristics of the films are improved for those grown using an oxygen flow rate of 0.6 L min^{-1} at 525 °C and 550 °C, which is related with the presence of denser and highly crystalline thin films. This is supported by the SEM images, which indicate that at higher substrate temperatures the grain sizes are larger and well-packed in comparison with the structures formed at lower temperatures.

Table 4.2 illustrates the calculated n and k of the films deposited at different oxygen flow rates and substrate temperatures for the monoclinic (semiconducting) phase (below T_c) and the tetragonal (metallic) phase (above T_c) at wavelength of 1500 nm. It is seen that generally, n decreases and k increases when heating the sample from 25 °C (below T_c) to 65 °C (above T_c).

Despite this overall trend, it is noteworthy for the samples grown using oxygen flow rate of 0.2 L min^{-1}, the values of n change significantly over the different deposition conditions from 1.746-2.212 for the monoclinic phase to 1.236-1.627 for the tetragonal phase. Similar variations are observed for the k values with variations ranging from 0.451-0.676 for the monoclinic phase to 1.652-1.933 for the tetragonal phase as shown in table 4.2. Regarding those grown using oxygen flow rate of 0.6 L min^{-1}, the n values were considerably higher for the monoclinic phase only, ranging from 2.145-2.843, while for the tetragonal phase the values obtained are not dissimilar to those obtained in the presence of reduced oxygen levels, i.e. 1.116-1.638. Similar observations are seen for the k values with variations ranging from 0.612-0.818 for the monoclinic phase to 1.953-2.124 for the tetragonal phase as shown in table 4.2.

Table 4.2. n and k values of vanadium oxide films grown in an APCVD reactor, recorded at 1500 nm, at 25 °C (below T_c) and 65 °C (above T_c). Films were grown using oxygen flow rates of 0.2 and 0.6 L min^{-1} at 500, 525 and 550 °C

	Oxygen Flow Rate / L min^{-1}							
	0.2				0.6			
	n		k		n		k	
	25 °C	65 °C	25 °C	65 °C	25 °C	65 °C	25 °C	65 °C
500 °C	1.821	1.493	0.542	1.652	2.423	1.116	0.612	2.124
525 °C	2.212	1.627	0.676	1.933	2.843	1.638	0.818	2.017
550 °C	1.746	1.236	0.451	1.785	2.145	1.332	0.757	1.953

It is significant to note that the presence of higher levels of oxygen results in consistently higher k values, for all growth temperatures, such that it may be stated that in general, oxygen rich growth conditions serve to increase both the real and imaginary parts of k, both before and after the phase transition. The implication here is that denser, more compact, possibly more crystalline films result from this interaction.

The changes observed in the optical constants of the films during the phase transition can be explained in terms of the relationship between the optical constants and materials structure.

According to Maxwell's electromagnetic theory and Classical Mechanics, the relationship between optical properties and structure may be described by equation 4.1 [28] below

$$\frac{n^2-1}{n^2+2} x \frac{3\varepsilon_o}{N} = \frac{\alpha\rho}{M} \quad (4.1)$$

where, M is the molecular weight, ρ is the density, α is the polarizability, n is the refractive index, N and ε_o are Avogadro's number and the vacuum permittivity respectively.

The lengths of V-O bonds ($d_{V\text{-}O}$) are different, when [VO_6] is distorted in the octahedral crystal field present in the case of the lower temperature phase (below T_c). The values of $d_{V\text{-}O}$ between V^{4+} ions and O^{2-} ions in the monoclinic structure are 0.126, 0.186, 0.187, 0.201, 0.203 and 0.205 nm [28]. In this case, polarization occurs when $d_{V\text{-}O}$ is less than the sum of the radii of the V^{4+} ion and O^{2-} ion ($R_{V\text{-}O} = r_{V4+} + r_{O2-} = 0.200$ nm). The greater the difference between $d_{V\text{-}O}$ and $R_{V\text{-}O}$, the higher the polarization and the larger the polarizability will be. However, in the case of VO_2 at high temperatures (above T_c), [VO_6] becomes tetragonal in structure and $d_{V\text{-}O}$ equals 0.194 nm, very close to the ideal value for $R_{V\text{-}O}$ [28]. Therefore, the polarizability of the tetragonal form is less than that for the monoclinic form. As the ion accumulation density is similar for these two different structures, the result of the reduction in polarizability is that n must also decrease. Thus, the n value of VO_2 decreases when VO_2

changes from monoclinic structure (below T_c) to tetragonal rutile structure above T_c due to the thermally induced phase transition [28].

This approach adequately explains the general trend observed, but cannot account for the observed variation in both n and k as a function of growth conditions. However, this is a particularly significant finding for the design of a CVD process for the production of high-quality thermochromic VO_2 thin films.

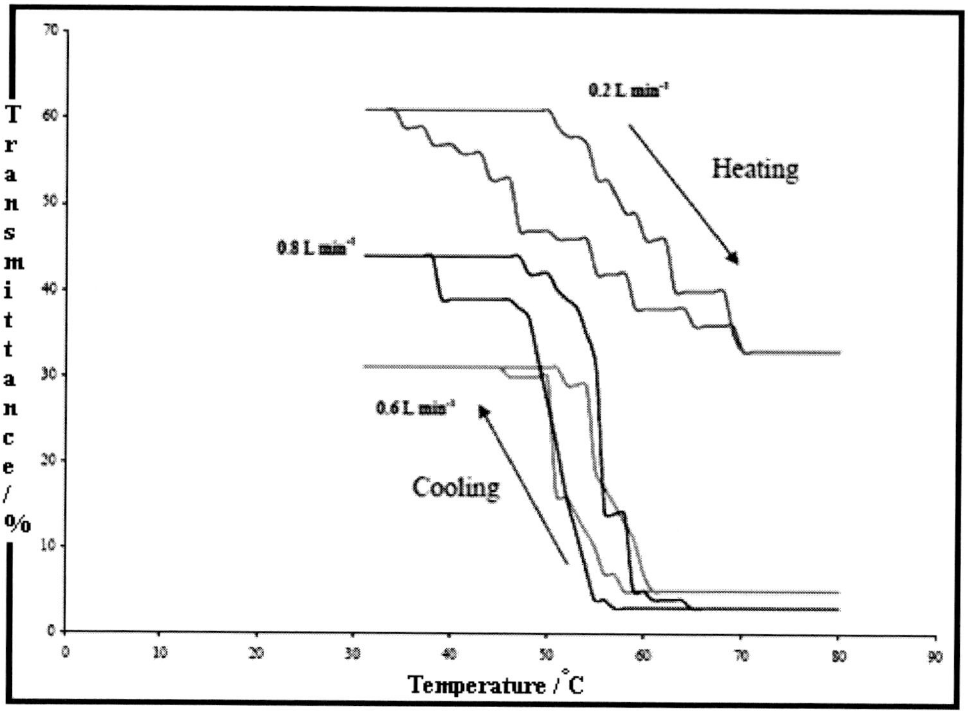

Figure 4.5. Transmittance at 2.5 μm against temperature for vanadium dioxide films grown using oxygen flow rates of 0.2 L min^{-1}, 0.6 L min^{-1} and 0.8 L min^{-1} at 525 °C. The direction of cycling is also indicated.

The transition temperature for the films grown at 525 °C for different oxygen flow rates was estimated by studying the change in transmittance at 2.5 μm as shown in figure 4.5. The T_c was taken to be at the centre of the hysteresis loop, while the width of the loop was also recorded as being of interest in terms of defining the sharpness of the transition. The T_c values and the hysteresis loop widths were recorded for oxygen flow rates of 0.2 L min^{-1}, 0.6 L min^{-1} and 0.8 L min^{-1}, being 54.5 °C and 9 °C, 54 °C and 5 °C, 51.5 °C and 5 °C respectively. From figure 4.5, it is again apparent that the oxygen flow rate is a quite crucial parameter in terms of obtaining thermochromic VO_2 thin films, in this instance as measured by the presence of narrow phase transition hysteresis loop widths and low T_c.

Figures 4.6 (a) and (b) address the effect of the growth temperature upon the thermochromic parameters of the films. It is clearly seen that the lowest deposition temperature (500 °C) shows the widest hysteresis loop and the highest transition temperature, which are 12 °C and 57 °C respectively. On the other hand, the narrowest hysteresis loop width was observed at 550 °C, which was 5 °C with T_c of 53.5 °C.

A possible reason for this variation in T_c and hysteresis loop width is the variation in the degree of crystallization and the morphology of the films induced by different deposition temperatures, in the presence of sufficient oxygen to get VO_2. Thus, growth at higher temperatures may simply increase the amount of crystalline VO_2 as compared to amorphous material, or it may result in growth of material with a different grain boundary structure. This particular point is supported in part by SEM images, which reveal a variation in microstructure with growth temperature. Since the phase transition involves physical movement of atoms within the lattice, it would not be surprising to find that the grain boundary influenced this process considerably.

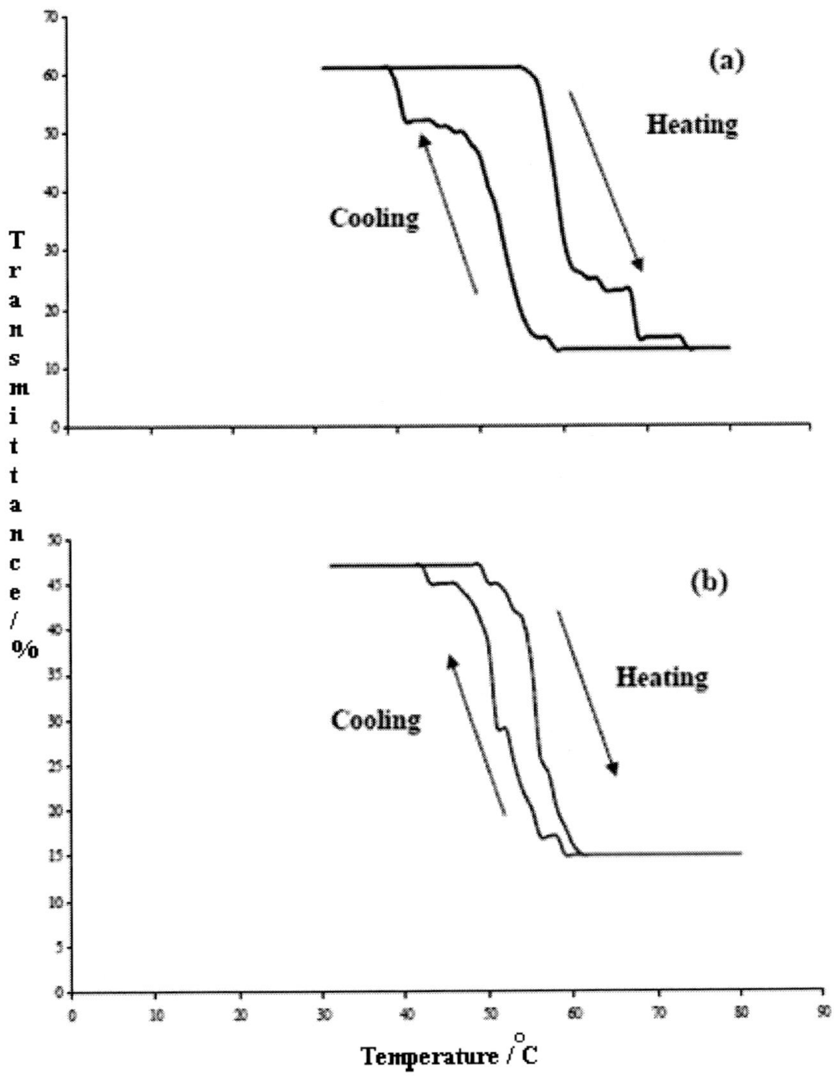

Figure 4.6. Transmittance at 2.5 μm against temperature for vanadium oxide films grown at 500 °C (a) and 550 °C (b), using an oxygen flow rate of 0.8 L min^{-1}.

5. IN-SITU MONITORING OF THE $VO(ACAC)_2$ CVD PROCESS BY FTIR WITH AND WITHOUT THE PRESENCE OF O_2

IR of $M(acac)_2^-$ and $M(acac)_3^-$ type complexes have been studied extensively. Theoretical band assignments were first made by Nakamoto and Matell [29], who carried out normal coordinate analysis on the 1:1 model of $Cu(acac)_2$. In addition, Mikami et al. [30] performed normal coordinate analysis on the 1:2 (square-planar) and 1:3 (octahedral) models of various acac complexes. On the other hand, the $VO(acac)_2$ has been examined first time in our laboratory [17].

Figure 5.1 depicts IR spectra of the gas phase for the reactive species obtained at different substrate temperatures over the range of 200-600 °C, without the presence of oxygen. Other researchers have also investigated the thermal decomposition of metal acetylacetonate compounds i.e. $La(acac)_3$ [31], $Sm(acac)_3$ [32], $Zn(acac)_2$ [32], $Fe(acac)_3$ [32], $Zr(acac)_4$ [33] and have found that the acac ligand decomposes to give a metal acetate with propyne initially, and then the acetate breaks down to yield metal oxycarbonate and C_3H_6O. However, the spectra depicted in figure 5.1 are quite complicated. Hence, a method was devised to analyse them in a systematic manner.

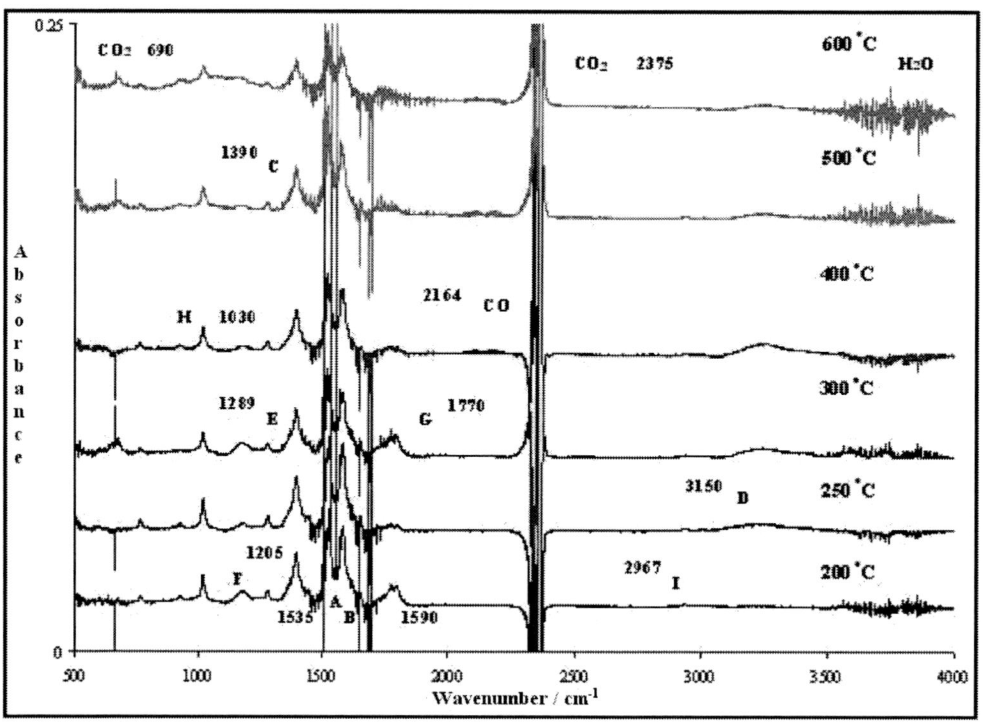

Figure 5.1. IR spectra of $VO(acac)_2$ system at 200 °C, 250 °C, 300 °C, 400 °C, 500 °C and 600 °C with 0 L min^{-1} oxygen flow rate and a total gas-flow rate of 12 L min^{-1} for the mid-IR region of 500–4000 cm^{-1}.

This method involved the label of each peak, the determination of peak area as a function of temperature, and then the plot of the data in order to correlate the trend observed with those trends obtained for other peaks in the spectra [17]. In general, it was observed that all bands have the same basic trend in that over the range of 200–300 °C, the area under the peaks increases, and then above this temperature decreases.

Bands A, B, C, E, F and H may be attributed to either unreacted $VO(acac)_2$ precursor or to species very similar in structure to the acetylacetonate ligand that contain the softened C–O bonds associated with this type of ligand operating in a bidentate fashion. In this respect, we cannot determine whether or not the V atom is still present because by comparison with the spectrum of the solid precursor [17], it is apparent that the V=O band would be too weak to be observed in figure 5.1, even if the precursor was intact. These bands are similar to those observed by other researchers, who have assigned these features to ethanoate species [31-34].

Strong OH stretching bands due to H_2O [35] and CO_2 [36] stretching bands were seen at all substrate temperatures as shown in figure 5.1. Regarding CO_2, it is observed that some of the peaks have different direction than the rest on the spectra at 200 °C, 250 °C and 400 °C at 690 cm^{-1} and 2375 cm^{-1}. Again it is believed that these are caused by small fluctuations in humidity in the external gas environment, which the IR beam passes through, leading to miscancellations between background and sample spectra. It is therefore impossible to ascertain whether H_2O and CO_2 are detected as decomposition products.

The only new bands detected (apart from those attributed to the miscancellation of bands due to atmospheric gases) are bands G and D. Band G lies in the spectral region associated with a carbonyl species, usually one found in acid anhydride species. Band D would also be consistent with this assignment since it would be assigned to the characteristic broad C–H stretching modes associated with species such as $(CH_3CO)_2O$, although the behaviour of this band with increasing temperature does not appear to be exactly the same as that observed for band G. In this respect it is noteworthy that band D appears most prominently in the spectrum recorded at 400 °C. This spectrum also shows the appearance of a band centred near 2145 cm^{-1}. This band is assigned to the formation of carbon monoxide (CO) [37] resulting from the decomposition of either the $VO(acac)_2$ itself or the decomposition of possible products arising from the original precursor such as the ethanoate species or the anhydride species.

Given the uncertainty in the band assignments, it is not possible to define a mechanism for the decomposition of $VO(acac)_2$ to VO_2 based solely on the spectra obtained. However, given that the only bands observed in the spectra that cannot be explained in terms of the vibrations of the original precursor would be consistent with the formation of an acid anhydride, it is possible to speculate the following possible decomposition routes (1)-(3) [17]:

$$VO(C_5H_7O_2)_2 \rightarrow VO(CH_3COO)(C_5H_7O_2) + C_3H_4 \quad (1)$$

$$VO(CH_3COO)(C_5H_7O_2) \rightarrow VO(CH_3COO)_2 + C_3H_4 \quad (2)$$

$$VO(CH_3COO)_2 \rightarrow (CH_3CO)_2O + VO_2 \quad (3)$$

Figure 5.2 depicts the IR spectra of the gas phase recorded from the CVD reactor in the presence of $VO(acac)_2$ with the addition of oxygen (0.2 L min^{-1}) as a function of temperature over the range of 200–600 °C. The spectra obtained are clearly very similar to those in the absence of oxygen (figure 5.1). In addition, the area under the peaks decreases with increasing temperature in a similar manner as above.

Bands A, B and C have been assigned to acetate species as already observed by other researchers [31-34]. Band E represents a C-C deformation of an unsaturated aliphatic unit, i.e. sp^3 hybridised C-C bond. While, band F is correlated with the asymmetric C-C and C-O bonds attached to a C=O. Regarding the bands G, H and I are allocated as C=O, asymmetric O-C and C-O bonds and asymmetric $–CH_3$ correspondingly. Bands E, F, G, H and I are most likely associated with the formation of an ester [38], possibly an unsaturated one, although, it is noteworthy here that the band assigned to an unsaturated C-H stretch seen in figure 5.1 is not present in figure 5.2.

It is appropriate to compare the spectra obtained for the decomposition of $VO(acac)_2$ with and without the presence of oxygen, since it was established on the previous section that addition of oxygen was necessary to induce the vanadium oxide growth.

From figures 5.1 and 5.2, it may be seen that in the presence of oxygen, the C-H stretching vibrations at 3000-3300 cm^{-1} are much less pronounced. Hence, it is suggested that in the presence of oxygen, the unsaturated species of sp^2 or sp C-H may not form. It is also noteworthy that in the presence of oxygen, the band due to CO [37] formation appears more pronounced. Thus, it seems that oxygen promotes the formation of CO and possibly reduces the stability of the unsaturated ester. Therefore, the proposed mechanism shown in the reaction above may not be valid for this system because oxygen may coordinate to the vacant octahedral site around the V^{4+} ion and weaken the other V-O bonds accordingly.

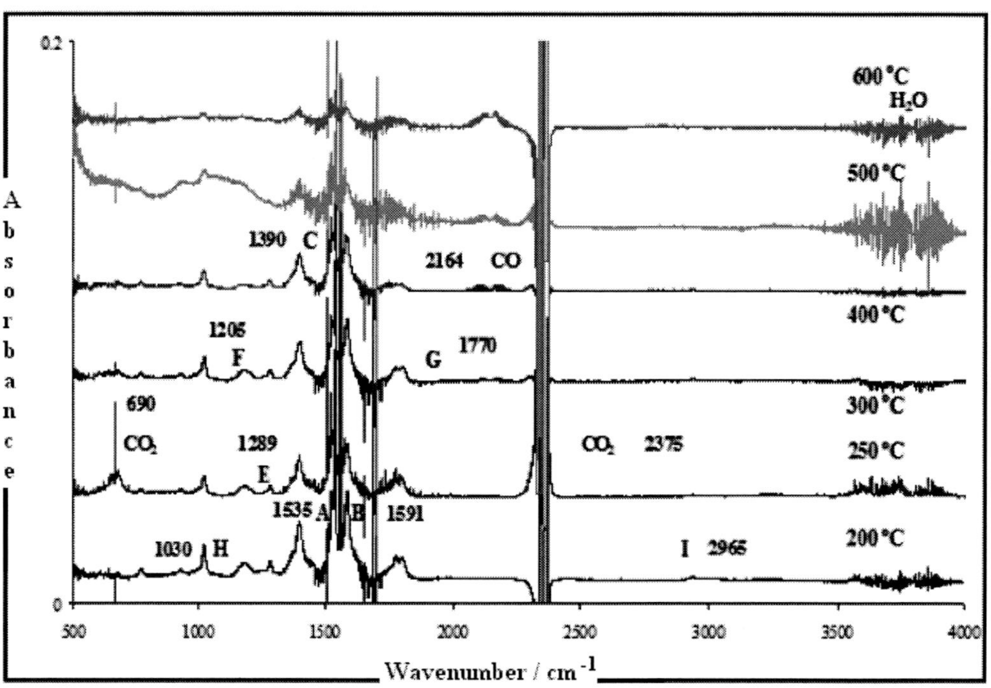

Figure 5.2. IR spectra of $VO(acac)_2$ system at 200 °C, 250 °C, 300 °C, 400 °C, 500 °C and 600 °C with 0.2 L min^{-1} oxygen flow rate and a total gas-flow rate of 12 L min^{-1} for the mid-IR region of 500–4000 cm^{-1}.

Carbon dioxide [36] bands and strong OH stretching bands due to H_2O [35] were also observed at all substrate temperatures. It is seen that some of the peaks have different direction than the rest at 200 °C, 400 °C, 500 °C and 600 °C at 690 cm^{-1} and 2375 cm^{-1}. These are caused by miscancellations between background and sample spectra. It is therefore impossible to ascertain whether H_2O and CO_2 are detected as decomposition products as already mentioned above.

In summary, the spectra obtained with and without the presence of oxygen suggest that at least one intermediate species in the decomposition of the $VO(acac)_2$ precursor is some form of ester, possible one which contains an sp^2 or sp CH, although, these latter species could arise from the direct formation of propyne as has been postulated in the decomposition of other metal acac complexes [31-34]. Thus, in this respect the decomposition of $VO(acac)_2$ seems to be rather different from the behaviour observed for other metal acac compounds.

The addition of oxygen suppresses the C-H region of the spectrum overall and results in an enhancement in the production of CO, while the rest of the spectrum is largely unchanged. It is therefore possible that in the presence of oxygen some oxidation of reaction products occurs possibly catalysed by the VO_2 deposits themselves [39, 40].

CONCLUSION

The growth of thermochromic vanadium dioxide films is possible by atmospheric pressure CVD using $VO(acac)_2$ as precursor. Conditions have clearly been identified within which single phase vanadium dioxide is formed.

The films presented a significant reduction in switching temperature (51.5 °C) in comparison with that obtained for vanadium dioxide grown using VCl_4 and H_2O (58 °C) [16], without degradation of hysteresis loop width (ca. 5 °C) and with improved switching contrast (28.8 and 36.1 % for transmittance and reflectance respectively). It is proposed that film crystallinity is the key parameter in enhancing the thermochromic properties with the presence of other phases or even amorphous material resulting in an increase in T_c, a reduction in switching properties and a general degradation of thermochromic properties, which is in agreement with other researchers [41].

The gas phase chemistry of the VO_2 formation from $VO(acac)_2$ with and without the presence of oxygen has been studied using in-situ FTIR during the APCVD reactions. A possible mechanism for the formation has been proposed. This involves as a first step the elimination of C_3H_4 to yield an ester species, possibly $VO(CH_3COO)_2$, which decomposes to yield an acid anhydride, most likely $(CH_3CO)_2O$ and VO_2.

This chapter has extensively studied the optimization of the APCVD processes using $VO(acac)_2$ to get enhanced switching characteristics. The fundamental chemical mechanisms underlying the particular APCVD processes have been examined to gain an insight into the film growth.

These findings are very important because they are likely to play an important role in influencing film growth rates and other film characteristics for smart window and other related applications.

ACKNOWLEDGMENTS

The authors would like to thank the EPSRC with reference number GR/R03020/01 for financial support, Pilkington Glass, UK for supply of the glass substrates, Dr Martin Davies and Dr Tom Paterson for the set up of the FTIR in-situ monitoring system as well as for the useful discussions.

REFERENCES

[1] http://www.azom.com/details.asp?ArticleID =154.
[2] F.J. Morin, *Phys. Rev. Lett.* 3 (1959) 34.
[3] G.A. Rozgonyi, W. J. Polito, *J. Electrochem. Soc.* 1 (1968) 576.
[4] F. Béteille, J. Livage, *J. Sol-Gel Sci. Technol.* 13 (1998) 915.
[5] C. Leroux, G. Nihoul, G. Van Tendeloo, *Phys. Rev. B* 57 (1998) 5111.
[6] J.B. Goodenough, *J. Solid State Chem.* 3 (1971) 490.
[7] D. Barreca, L.E. Depero, E. Franzato, G.A. Rizzi, L. Sangaletti, E. Tondello, U. Vettori, *J. Electrochem. Soc.* 146 (1999) 551.
[8] J.B. MacChesney, H. J. Guggenheim, *J. Phys. Chem. Solids* 30 (1969) 225.
[9] T.E. Phillips, R.A. Murray, T.O. Poehler, *Mater. Res. Bull.* 22 (1987) 1113.
[10] J.W. Pierce, J.B. Goodenough, *Phys. Rev. B.* 5 (1972) 4104.
[11] M.N. Field, I.P. Parkin, *J. Mater. Chem.* 10 (2000) 1863.
[12] T.D. Manning, I.P. Parkin, *Polyhedron* 23 (2004) 3087.
[13] Y. Takahshi, M. Kanamori, H. Hashimoto, Y. Moritani, Y. Masuda, *J. Mater. Sci.* 24 (1989) 192.
[14] T. Maruyama, Y. Ikuta, J. Mater. Sci. 28 (1993) 5073.
[15] M.B. Sahana, G.N. Subbanna, S.A. Shivashankar, *J. Appl. Phys.* 92 (2002) 6495.
[16] D. Vernardou, M.E. Pemble, D.W. Sheel, *Chem. Vap. Deposition* 12 (2006) 263.
[17] D. Vernardou, M.E. Pemble, D.W. Sheel, *Thin Solid Films* 516 (2008) 4502.
[18] J. Cui, D. Da, W. Jiang, *Appl. Surf.* Sci. 133 (1998) 225.
[19] M. Ghanashyam Krishna, Y. Debauge, A. K. Bhattacharya, *Thin Solid Films* 312 (1998) 116.
[20] J. Mendialdua, R. Casanova, Y. Barbaux, *J. Electron Spectrosc. Relat. Phenom.* 71 (1995) 249.
[21] E. Cazzanelli, G. Mariotto, S. Passerini, W. H. Smyrl, A. Gorenstein, *Sol. Energy ater. Sol.* C 56 (1999) 249.
[22] M. Bennoussa, E. Ibnouelghazi, A. Bennouma, E.L. Ameziane, *Thin Solid Films* 265 (1995) 22.
[23] T.D. Manning, I.P. Parkin, M.E. Pemble, D.W. Sheel, D. Vernardou, *Chem. Mater.* 16 (2004) 744.
[24] K.D. Rogers, *Powder Diffr.* 8 (1993) 240.
[25] J.C. Parker, *Phys. Rev. B* 42 (1990) 3164.
[26] C.K. Carniglia, K.N. Schrader, P.A. O' Connel, S.R. Tuenge, *Appl. Opt.* 28 (1989) 2902.
[27] Pro-optix User Manual, Version 2.0, *Aquila Instruments* Ltd (2000).

[28] S. Lu, L. Hou, F. Gan, *J. Mater. Sc.* 28 (1993) 2169.
[29] K. Nakamoto, A.E. Martell, *J. Chem. Phys.* 32 (1960) 588.
[30] M. Mikami, I. Nakagawa, T. Shimanouchi, Spectrochim. *Acta 23A* (1967)1037.
[31] G.A.M. Hussein, H.M. Ismail, *Powder Technol.* 84 (1995) 185.
[32] H.M. Ismail, *J. Anal. and Appl. Pyrol.* 21 (1991) 315.
[33] H.M. Ismail, *Powder Technol.* 85 (1995) 253.
[34] H.M. Ismail, *Colloids Surf.* A 97 (1995) 247.
[35] C.N. Banwell, Fundamentals of Molecular Spectroscopy, 2nd Edn. McGraw-Hill Book Company Ltd, London, 1972, pp.88.
[36] C.N. Banwell, Fundamentals of Molecular Spectroscopy, 2nd Edn. McGraw-Hill Book Company Ltd, London, 1972, pp.90.
[37] C.N. Banwell, Fundamentals of Molecular Spectroscopy, 2nd Edn. McGraw-Hill Book Company Ltd, London, 1972, pp.81.
[38] B. Smith, Infrared Spectral Interpretation "*A Systematic Interpretation*", CRS Press, Florida, 1999, pp. 110.
[39] G.I.N. Waterhouse, G.A. Bowmaker, J.B. Metson, *Appl. Catal. A General* 266 (2004) 257.
[40] P. Forzatti, E. Tronconi, A.S. Elmi, G. Busca, *Appl. Catal. A General* 157 (1997) 387.
[41] E.E. Chain, *Appl. Opt.* 30 (1991) 2782.

In: Vanadium
Editors: V. N. Baranova and A. V. Fortunatov

ISBN 978-1-62081-600-4

Chapter 5

HOW MUCH VANADIUM IS THERE IN PAPER? A BIBLIOMETRIC OVERVIEW OF THE GLOBAL SCIENTIFIC PRODUCTION ON VANADIUM

Edoardo Magnone*
High Efficiency and Clean Energy Research Division,
Korea Institute of Energy Research (KIER), Daejeon, Republic of Korea

ABSTRACT

The chapter focuses on the progress made in the field of vanadium research over the last two decades. Many bibliometric studies have appeared in the scientific literature which focuses attention on the global trend for research productivity in the most varied subjects but at the moment there are no studies analyzing the world literature on vanadium. The purpose of this study is to identify research output, and research areas, as well as descriptively and quantitatively characterize the field of vanadium through publication trend analysis over the last twenty years.

The growth of the literature on vanadium was found exponential in nature. During 1990-2011 a total of 35594 outputs were published by scientists in this field in various countries. The average numbers of research output per year were 1618. United State is the top producing country with 6224 outputs (17.5%) followed by The People's Republic of China with 4599 outputs (12.9%) and Japan with 3672 outputs (10.3%). The Chinese absolute numbers of research outputs continued to increase over time. The rise of The People's Republic of China has been especially notable, overtaking United States and Japan in terms of its research output in recent years.

University of Tokyo topped the list with 375 research outputs (1.05%) followed by Russian Academy of Science with 349 outputs (0.98%) and CNRS Centre National de la Recherche Scientifique with 336 outputs (0.94%). About nine-tenths of the documents on vanadium were published in English language. The most preferred journals by the scientists in this topic were *Inorganic Chemistry* with 586 outputs (1.65%) followed by *Physical Review B Condensed Matter and Materials Physics* with 456 (1.28%) and

* Corresponding author: High Efficiency and Clean Energy Research Division, Korea Institute of Energy Research (KIER), 152 Gajeong-ro, Yuseung-Gu, Daejeon, 305-343, Republic of Korea, Tel.: +82-42-860-3384; fax: +82-42-860-3134, E-mail address: magnone.edoardo@gmail.com; magnone@kier.re.kr

Applied Catalysis A General with 444 outputs (1.25%). The most preferred subject area was Materials Science with 12925 outputs (36.31%). The high frequency keywords were Vanadium (14430 documents, 40.54%), Vanadium Compounds (6662 documents, 18.72%) and Vanadium Alloys (3741 documents, 10.51%). The top-five chemical elements in the keyword list were Chromium (2716 documents, 7.63%) followed by Titanium (2352 documents, 6.61%), Nickel (2331 documents, 6.55%), Aluminum (2184 documents, 6.13%) and Copper (2135 documents, 5.99%).

Keywords: Informetric methods, Academic Research Output, Publication, Vanadium

1. Introduction

In the last few decades there has been a growing interest for the development of vanadium for both fundamental and practical points of view in different fields of study [1-22]. Recently there has been a growing interest in the use of vanadium in renewable energy applications for batteries [16], high-performance supercapacitors [22], as therapeutic agents [5,10,13], *in the catalyst applications [6,14]* and for hydrogen purification membranes [23,24]

Despite increasing interest in the subject of this book, there have not been attempts at gathering systematic data on the global scientific production of vanadium research. Depending on the objectives, scientometric methods – a systematic measurement of science - could characterize the development of global scientific production in a specific research field [25, 26]. Many bibliometric studies have appeared in the scientific literature which focuses attention on the global trend for research productivity in the most varied subjects *but there are no studies* analyzing the world literature on vanadium [27].

In particular, the aim of present study was to quantitatively determine the growth and development of world literature on vanadium in terms of research output. In *the study reported in this chapter,* the author collected data from Scopus database [28]. The data so obtained was analyzed to study the world research activity on vanadium in general and to see the change in the pattern of research output in the time. This study is the first *scientometric* analysis in a field of growing *interest like vanadium.*

2. Data Collection and Methodology

It is a common opinion that the number of publications is directly proportional to scientific productivity on a subject. An analysis of the literature on vanadium - with particular emphasis on epistemology rather than chemical element vanadium - was performed on the basis of the keywords of the bibliographic database Scopus ® [28]. The number of research outputs depends on the scientific databases used but the Scopus's database selected is sufficiently representative of the progress of scientific knowledge on the vanadium fields. As described by the owners, Scopus is the largest abstract and citation database and contains over 19000 titles (18000 peer-reviewed journals, 400 trade publications and 300 book series) from more than 5000 international publishers in 45.5 million records (24.5 million records with references back to 1996, 21 million records pre-1996 which go back as far as 1823, 4.6

million conference papers from proceedings and journals) [29]. This database is representative of the overall situation [30, 31]. Article, review, letter, note, editorial, erratum, conference paper, conference review and book series in life (> 4,300 titles), health (> 6,800 titles), physical (> 7,200 titles) and social sciences & humanities (> 5,300 titles) were included in this analysis.

In this study, the metrics used for vanadium publication are numbers of research output [26]. One assumption made for this work is that the number of publications is directly proportional to scientific productivity in the world over the past twenty years. The search is confined to the publications from 1990 to 2011 and involved the extraction of papers with vanadium words in the Title, Abstract and list of Keywords fields and that were indexed in the Scopus ® database on January, 2011. Publication source is defined as the source of the research output, this field is labeled in Scopus as Article, Review, Conference Paper, Conference Review, Note, Letter, Short Survey, Erratum Editorial, Book, Report, Abstract Report, Business Article and Undefined.

Applying the above selected strategies the author got 35594 research output submitted in the target-time window. Collectively, the result gives us an approximate picture of the overall importance for the vanadium chemical element during last two decades. To the author's knowledge at the time of writing this chapter, no study has been published focusing on the relative value of the research output for vanadium. Preliminary results of this work and some empirical notes are described in the next section of this chapter.

3. Results and Discussions

3.1. Publication Source

During 1990-2011 a total of 35594 outputs were published on vanadium by researchers in various countries. According with Scopus database, there were 2161 research outputs published in 2011 with the world vanadium in their Titles, Abstracts or list of Keywords. The average number of publications produced per year was 1618.

Figure 1 depicts the overall research output (1990-2010) by publication sources. The publication sources are defined in the second section of the chapter. All over the world, like any other discipline and subject of science, the Articles account for the maximum number of research outputs with 29749 records (83.58%) followed by Conference Paper and Review with 4236 (11.90%) and 879 publications (2.47%), respectively.

Figure 2 gives an illustration of annual growth in the field of vanadium. It can be clearly visualized from the figure that growth of the cumulative number of literature was very slow during 1990-1996 and it peaked during 1996-2004, indicating that research on vanadium received a major boost during this period. In particular, it can be noted that during the 1995 the number of reviews increase considerably both in absolute terms as also in relation to papers. The phenomenal increase in the number of reviews and conference papers is, however, an indication of growing interest in the subject.

Even though there are fluctuations in number of research outputs every year, a significant mathematical correlation was found between the yearly cumulative number of research outputs and the year (1990-2011), as displayed in Figure 2. The cumulative document result progression, Paper, Review, and Conference Paper was represented by four exponential equations. Therefore, it can be concluded that the number of academic outputs on the topic of vanadium is still growing at a high rate. Moreover, based on the mathematical model during 1990-2011, it can be calculate that the number of research output on the topic of vanadium in 2030 will be twice of outputs in 2010.

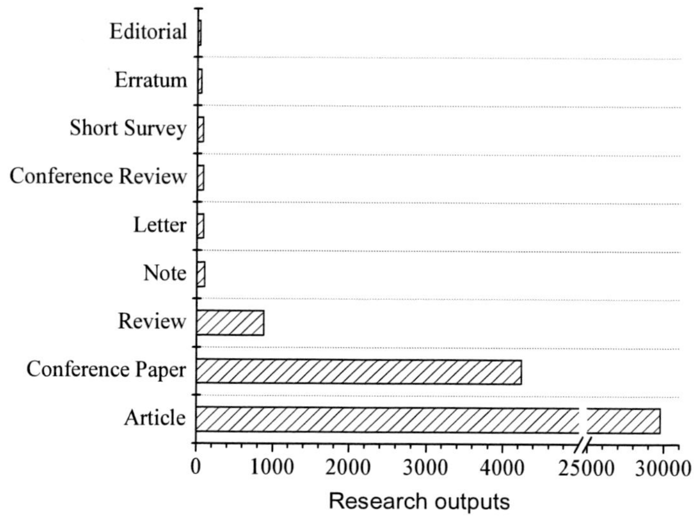

Figure 1. Publication source on vanadium all over the world (1990-2010).

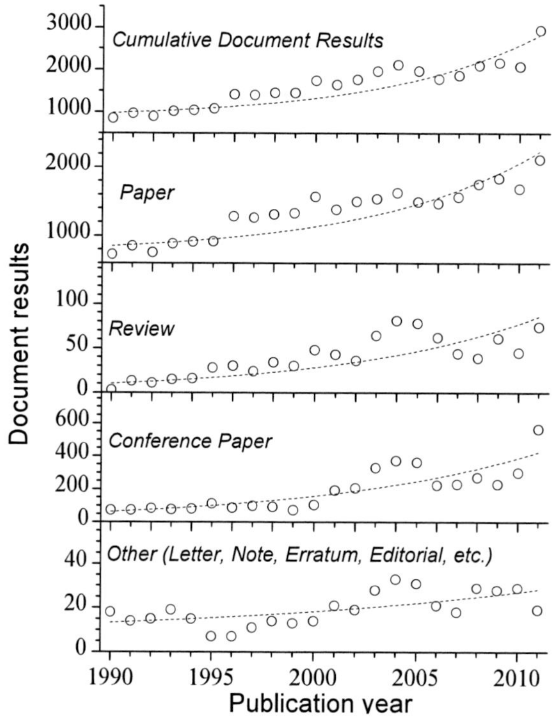

Figure 2. Trend in research output production by publication source.

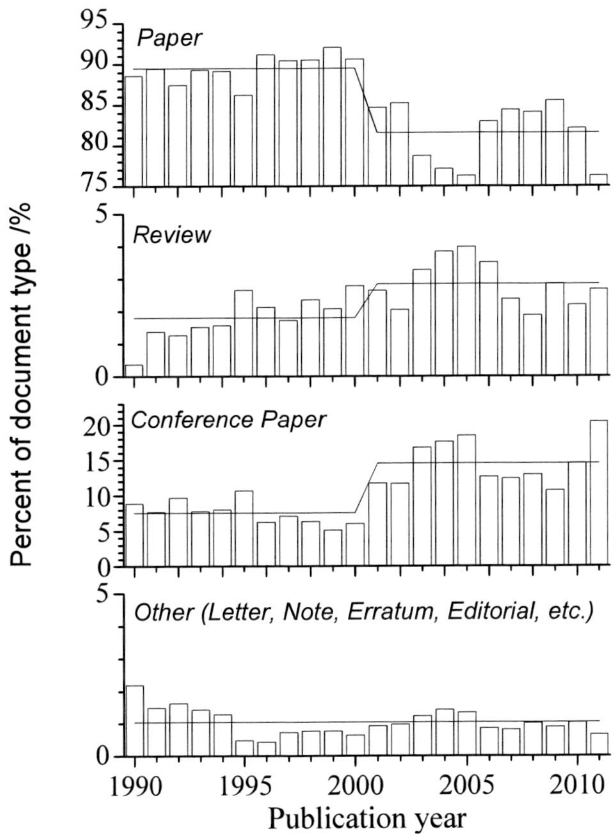

Figure 3. Proportion of publication source produced in the field between 1990 and 2011.

Figure 3 presents the percent of research output from 1990 to 2011 with their distribution by category of research sources. Just before the end of last century the average percent of Papers published in the international journals is about 89.5% (1990-2000), while the Conference Papers is only 7.60 % (1990-2000). In the second decades of analysis (2001-2011) the average percent of Conference Paper in the topic of vanadium increase to almost two times, while Reviews increased from 1.80 % in 1990-200 to 2.84 % in 2011. It is found that the sum of the percent of Letter, Note, Erratum, Editorial documents is essentially constant from year to year

3.2. Subject Categories

In the topic of vanadium, there was a great diversity on the subject areas identified by Scopus database during the last twenty years. All over the world, the six most common subject areas were Materials Science, Chemistry, Physics & Astronomy, Engineering, Chemical Engineering, and Biochemistry, Genetics & Molecular Biology. The total number of research output by major subject area-wise in the last 20 years is given in Figure 4. The total research output on vanadium related research from top five subject categories amounted to about 50%. More specifically, the proportion of research outputs per subject area exhibited some variation during the time period covered (Figure 5a). A distinct decrease of ranking appeared in Materials Science and Chemical Engineering areas. In the Materials Science field the percent of papers on vanadium decreased from 26.74% during 1990 to 22.16% in 2011.

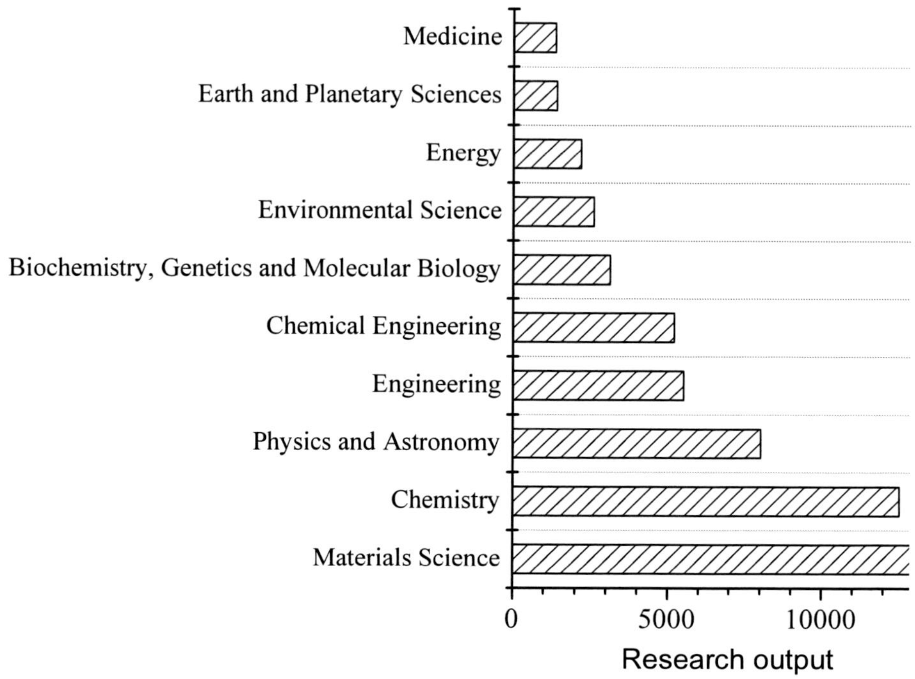

Figure 4. Subject category distribution of literature on the vanadium-related research.

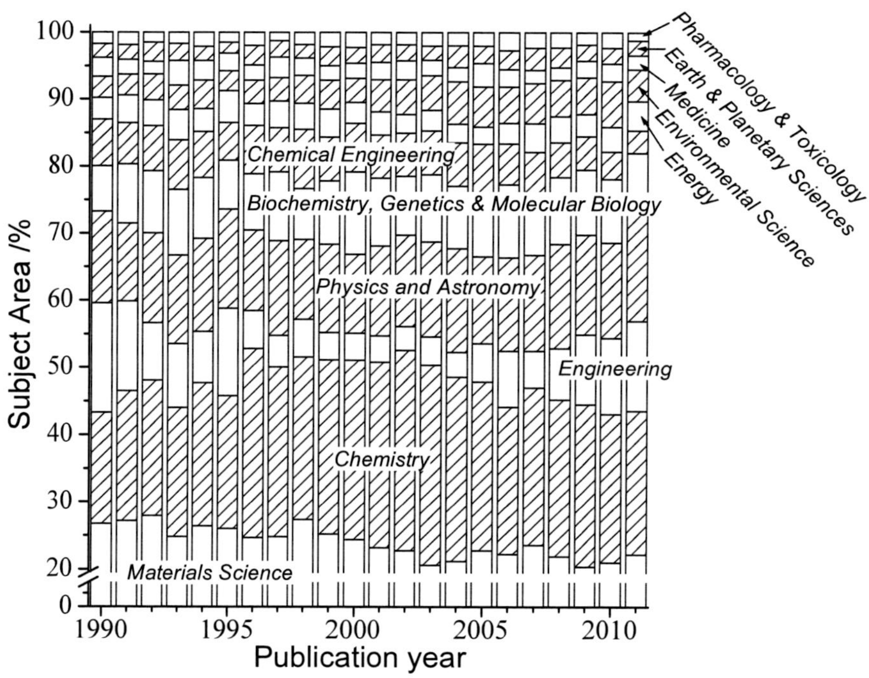

(a)

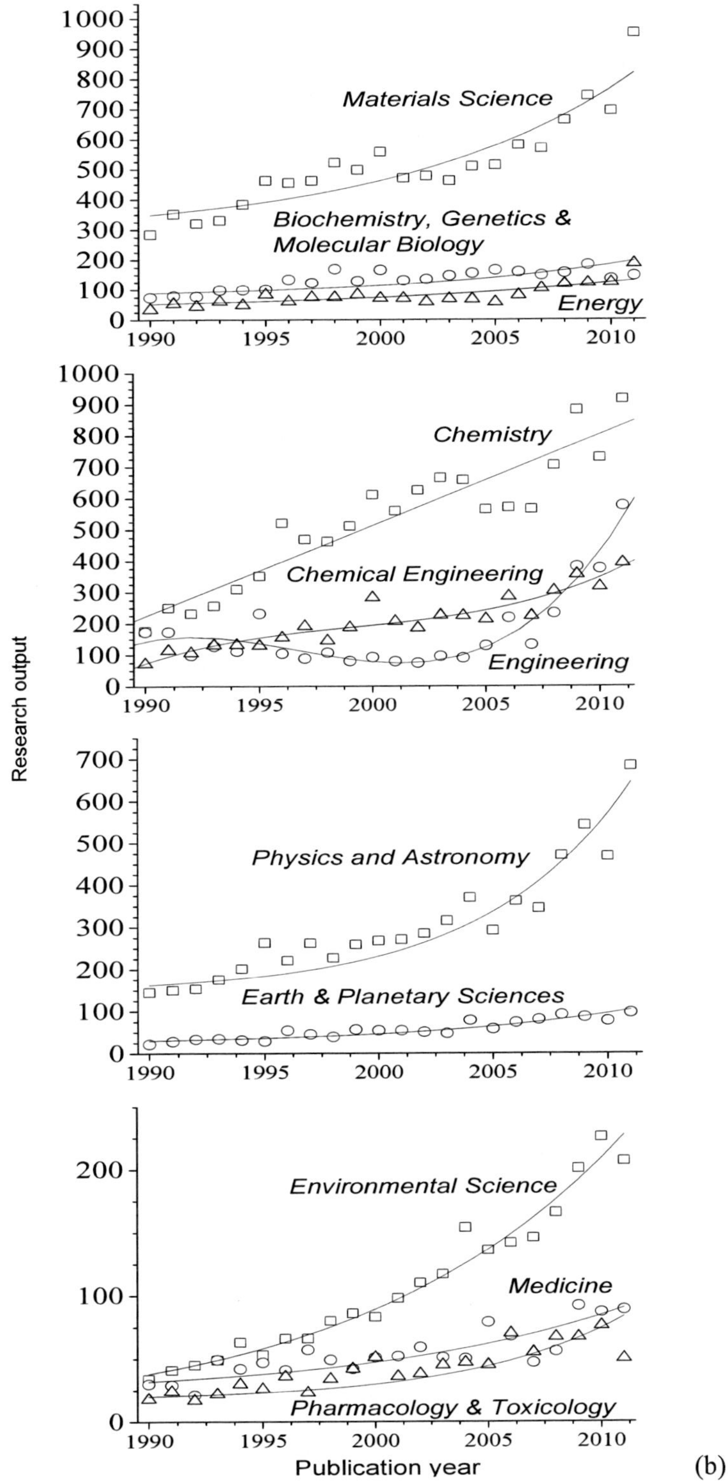

Figura 5. (a) Proportion of subject category distribution. (b) Growth in research output for selected subjects.

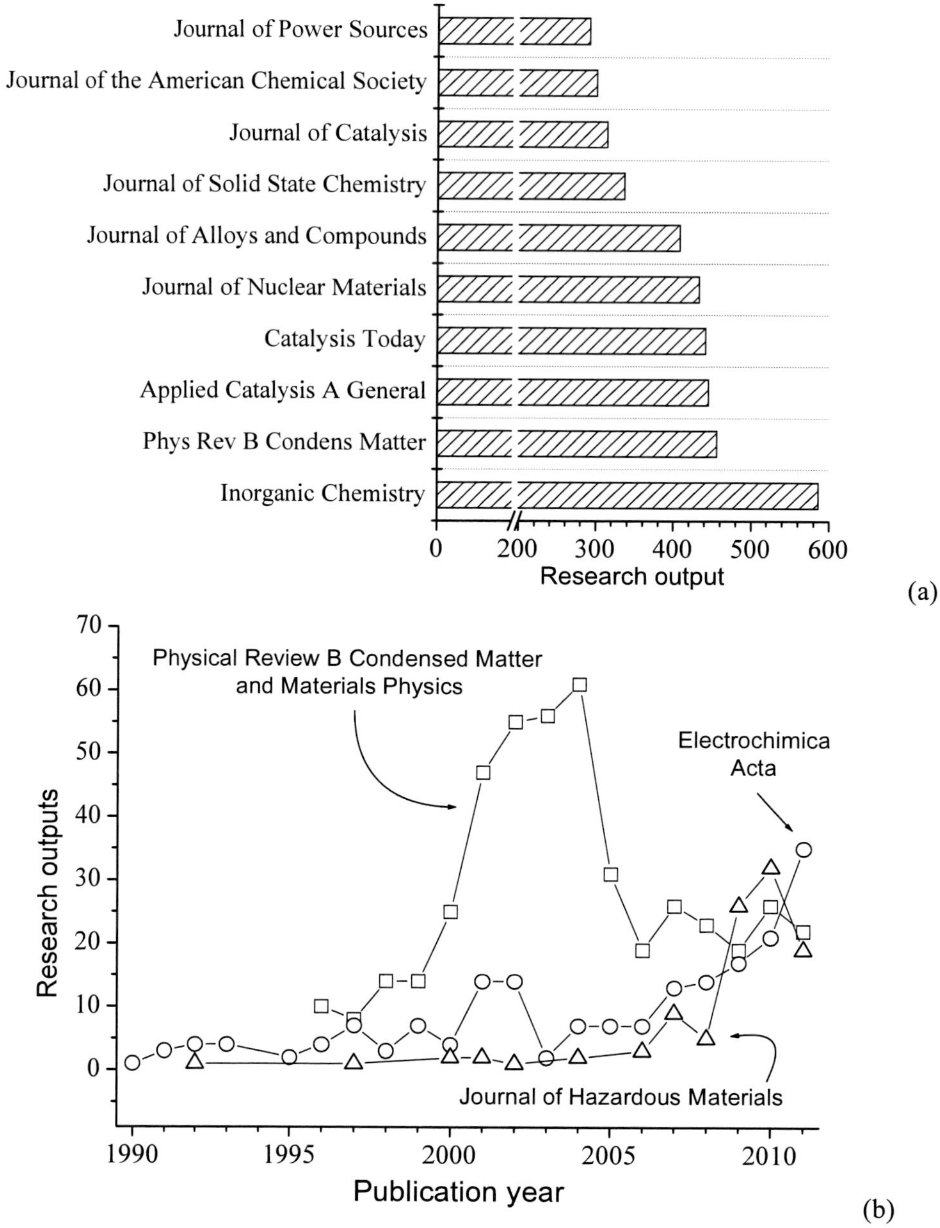

Figure 6. (a) Proportion of global research output by journal count. (b) Growth in research output for selected journals between 1990 and 2011.

Another important point is that as the Chemistry subject area increases from 1990 to about 2002, the percent of research output in the Engineering decreases. In fact, a completely different observation was made for Engineering. Over le last 20 years, this discipline showed an decrease in number of research outputs with the word vanadium in the Title, Abstract or list of Keywords.

Equally remarkable is that the Environmental Science and Pharmacology & Toxicology subject areas increased exponentially with time as shown in the Figure 5b. In comparison to the other subject areas investigated, however, it should first be noted that Environmental Science also began from a very low base line – for instance similar to the research outputs in the Medicine, Earth and Planetary Sciences, or Pharmacology and Tossicology in the year 1990 as a reference point - but the increase in vanadium papers in Environmental Science was bigger than for article in other fields. The large percent in this field indicates the growing interest in the environmental and human protection. The two subject areas are also significant because of their implications for global and national economies and politics.

3.4. Geographical Distribution of Research Output

A country's position in the international ranking can be estimated on the basis of some output indicators, where the number of research publications is the most frequently used output indicator [26,27]. The Figure 7a shows the geographic distribution of research output in the vanadium fields for the top ten countries. For a range of years between 1990 and 2011, 17.5% of research outputs were published by authors whose address was in United States (6224 outputs), 12.9% were published by authors in China (4599 outputs), and 10.3% were published in Japan (3672 outputs). Germany and Russian Federation accounts for about 7.13% (2538 outputs) and 6.34 % (2255 outputs), respectively, of the world production *in the same period of time*. The total research output on vanadium related research from top five countries amounted to about 50% and the remaining 50 percent is distributed in the other 131 countries.

Figure 7b presents the percent of research output from 1990 to 2011 with their distribution by *countries* in the research sources. It becomes clear that the United State's research output in this topic declined for the entire period. The United State accounts for about 30% of the world production of research output in 1994. In the same year, Japan accounts for about 17% of the world production. Japan share of the world production in the field of vanadium was more of 16% in 1991 and about 7% in 2010. The percent of outputs from The People's Republic of China increased strongly for the entire period. It should also be noted that the research output from the Republic of Korea show statistically increasing trends.

We examined the total number of records published by United States, People's Republic of China, Japan and South Korea, and then examined the annual average research outputs in the last twenty years. As Figure 7c shows, in 1995 the United States had approximately eight times the number of total research output of People's Republic of China but, by 2000-2005, the United States appeared to have peaked, while People's Republic of China still exhibited rapid growth for the entire period.

From 1996 to 2011, the Korea trend line was similar in shape, with People's Republic of China exhibiting a somewhat faster increase in growth. Japan is showing stagnation in number of research output in this field. Based on the exponential model during 1990-2011, it can be assumed that the number of research outputs on the topic of vanadium from South Korea will be surpass Japan in 2020.

The relative positions of The People's Republic of China, United State, and South Korea have been studied extensively over the last few years (see for instance Ref. [32]) and numerous reviews have appeared on this subject [33]. Generally speaking, China and South Korea are often cited as rising powers in science [34]. *In particular, in this chapter it* has been demonstrated that recently China has greatly increased it scientific activities in the vanadium-related field. The results are consistent also with the interpretation proposed by Leydesdorff and Zhou [35]. The author correctly argues that "*our center of gravity of the world system of science may be changing accordingly*" [35].

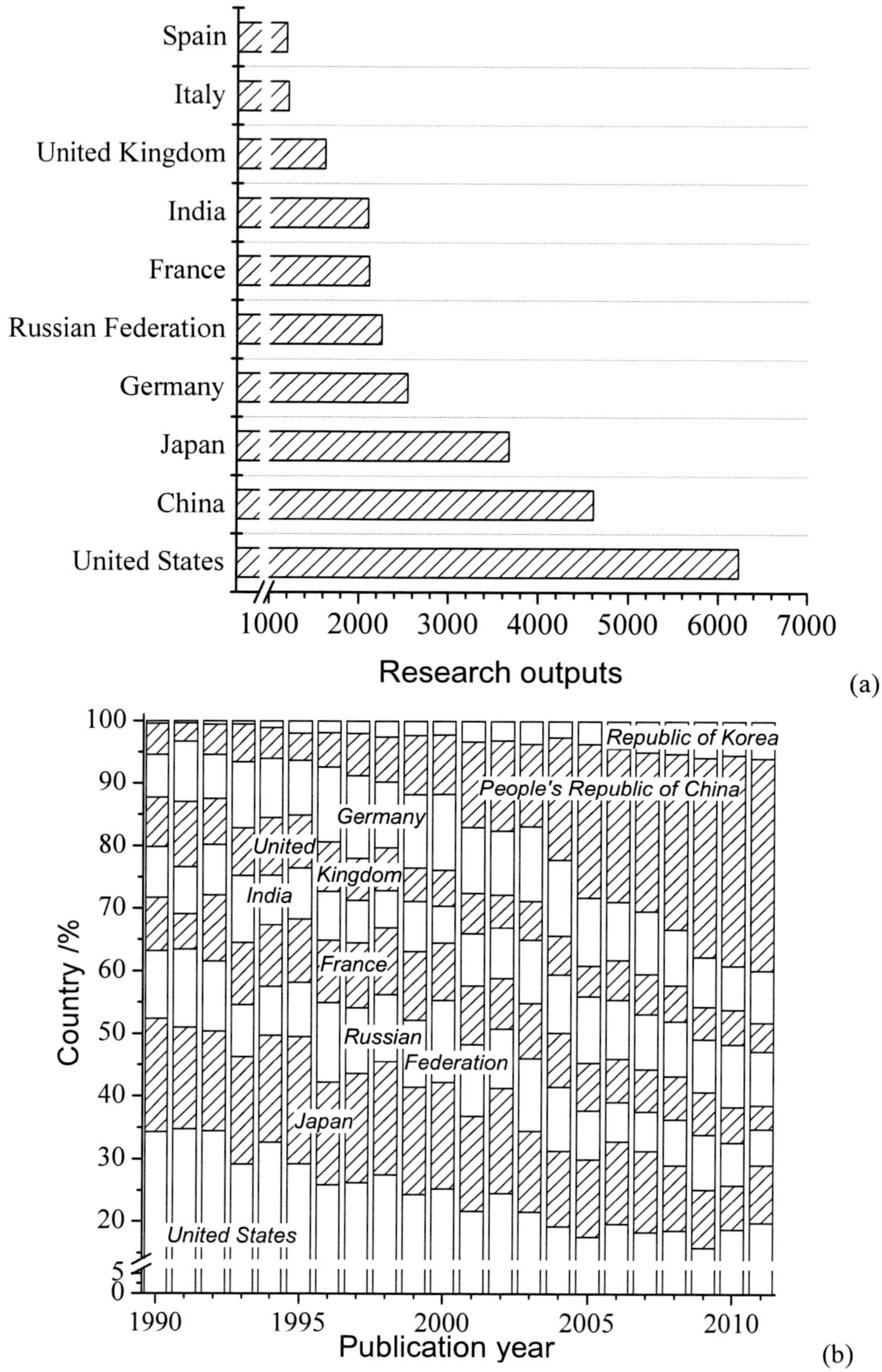

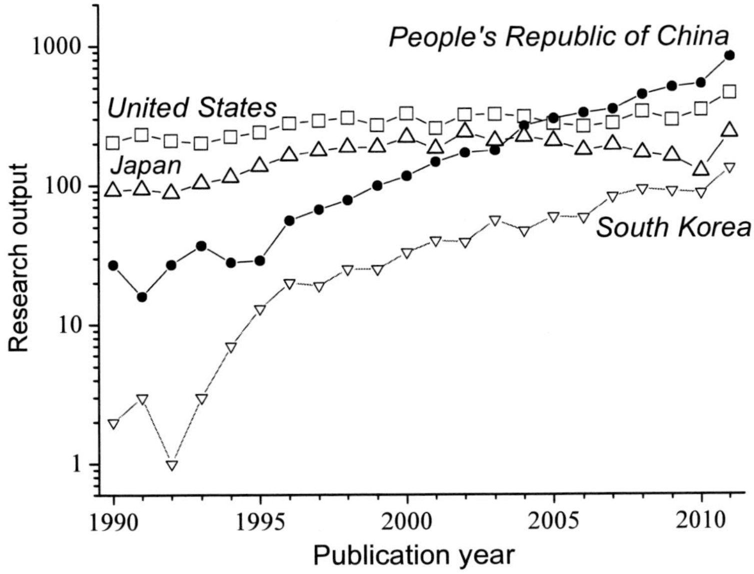

(c)

Figure 7. (a) Proportion of global research output by country. (b) Proportion of national publication output produced between 1990 and 2011. (c) Growth in research output for selected countries.

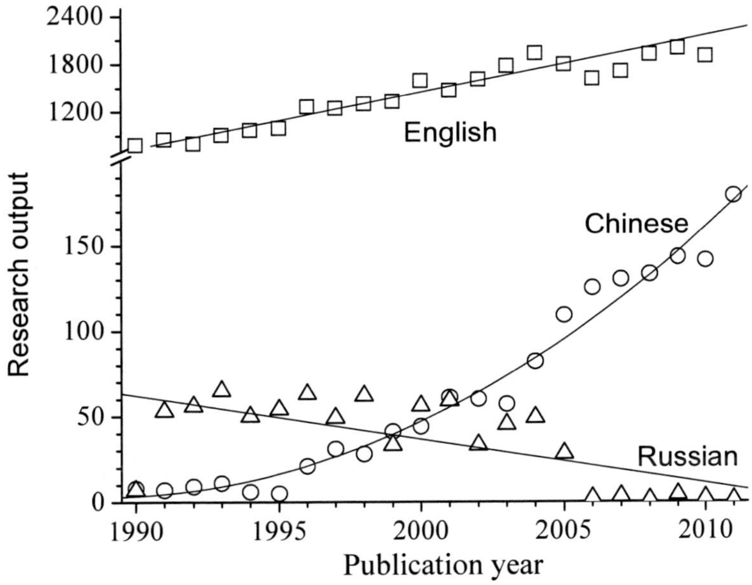

Figure 8. Increase in the time of the world's papers wrote in English, Chinese and Russian languages.

3.5. Language-Wise Distribution of Research Output

All over the world, 91.33 % of research outputs during the recent twenty years period (1990-2011) were published in English language. 4.01% was published in Chinese and 2.1%

in Russian language. About nine-tenths of the outputs on vanadium are written in English language, which indicates that English is consolidated as the main language in scientific communication in the world - "*the lingua franca of research*" [33]. Moreover, *in the first approximation, we can conclude* that the language of research outputs is tied to the geographical distribution of research output (discussed in an earlier section in this chapter). In particular, an interpretation based on the fact that approximately 20% of vanadium in the world is produced from primary ore from sources in north-western China and eastern Russia is consistent with the results reported in this chapter.

Figure 8 shows how the number of articles has grown and how their language distribution across the world has changed in recent years. People's Republic of China has increased its publications in Chinese to the extent that it is now the second highest producer of research output in the world. People's Republic of China has replaced the Russian Federation in the top three, climbing from 3rd in 1990 to 2nd position between 2000 and 2011.

3.6. Organization-Wise Distribution of Publications

This section reviewed the research output data for the twenty-year period from early 1990 to the finishing of the last year (2011). During the entire study period (Figure 9a), the highest-productivity country Asian institutions were the University of Tokyo (Japan), which published 375 outputs, Kyoto University (Japan), which published 267 outputs and Tohoku University (Japan) which published 233 outputs on vanadium. Indeed, it is remarkable to note that the among the top-twenty are universities from Japan (7), Russian Federation (4), United States (3), France (2), Germany (1), Brazil (1), China (1) and Poland (1).

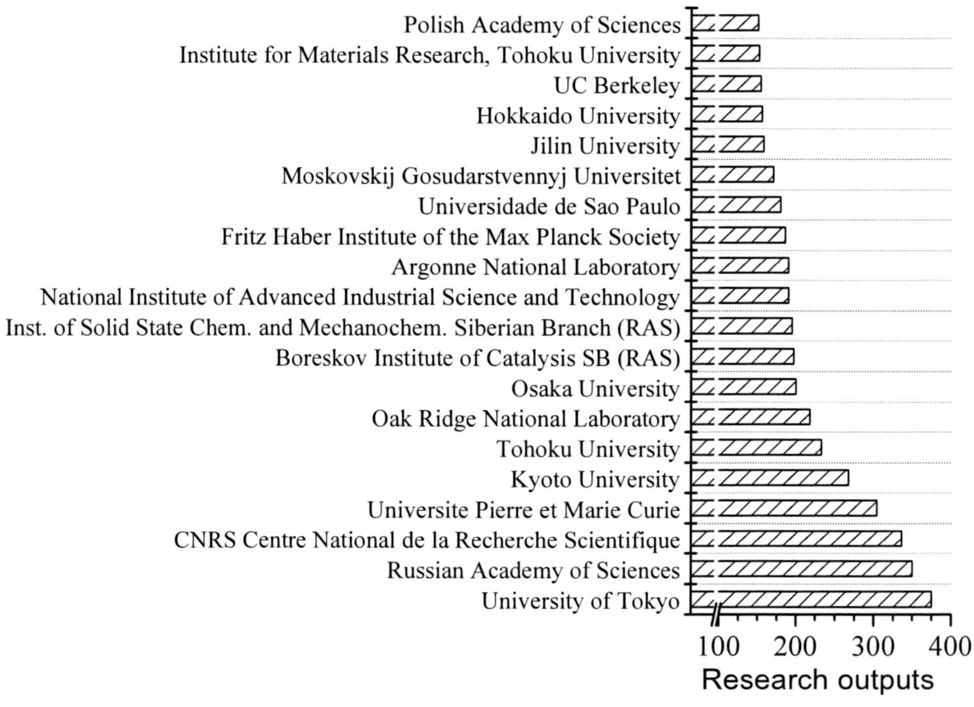

(a)

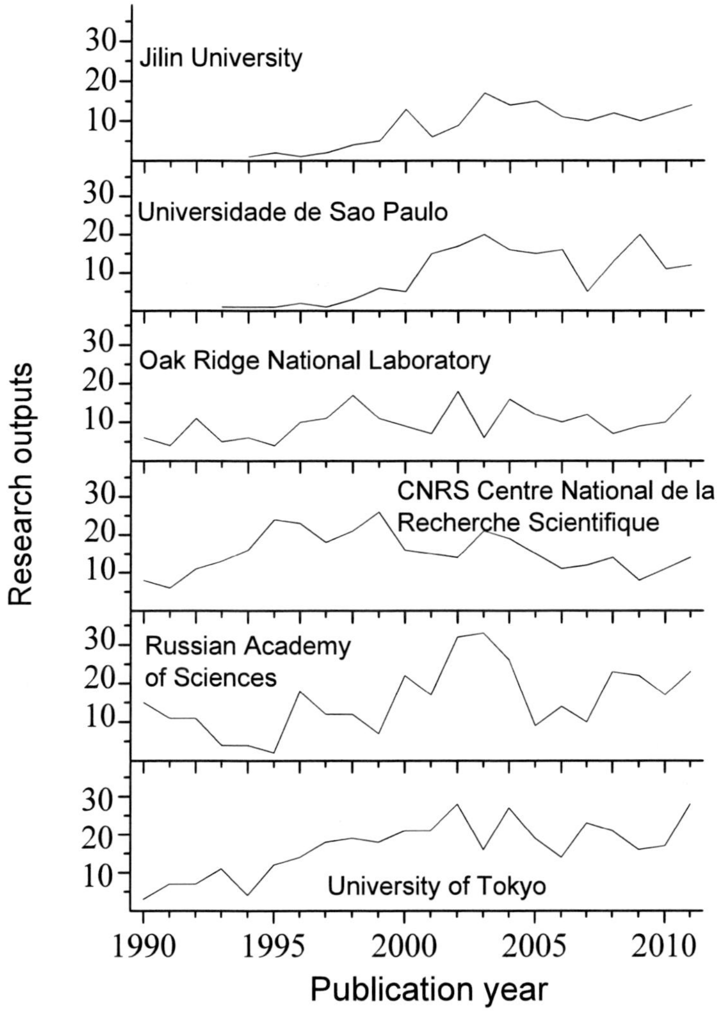

(b)

Figure 9. (a) Proportion of global research output by institution. (b) Growth in research output for selected institutions.

The results depicted in Figure 9b are representative of the performance comparison of the six institutions in different continents with similar research output levels. Vanadium publication outputs have been increasing rapidly in different institutions (such as University of Tokyo, Universidade de Sao Paulo and Jilin University). Figure 9b also demonstrates that the number of research outputs with Russian Academy of Science, CNRS Centre National de la Recherche Scientifique, and Oak Ridge National Laboratory affiliations has declined during the period of the study.

3.7. Most Prolific Authors

Scientific productivity of eminent scientists and the impact of research are essential in participatory research processes [26,27]. Furthermore, the objective here is not to create a ranking of scholars in vanadium subject. As the following graph shows (Figure 10), the most prolific authors were T. Murogata (Japan) who topped the list with 103 research outputs in the nuclear materials field and fusion technology [36], followed by D. Rehder (Germany) with 93 outputs in the subject area of Chemistry and Biochemistry [37], and J.Zubieta (United States) with 90 outputs about Materials Science, Chemistry, and Biochemistry subjects [4]. When interpreting these data, the reader of this book should keep in mind that maybe most of those who have had their work published in the topic of vanadium do not appear in Scopus database [30,31] and that the ranking is based only on data available on January, 2012.

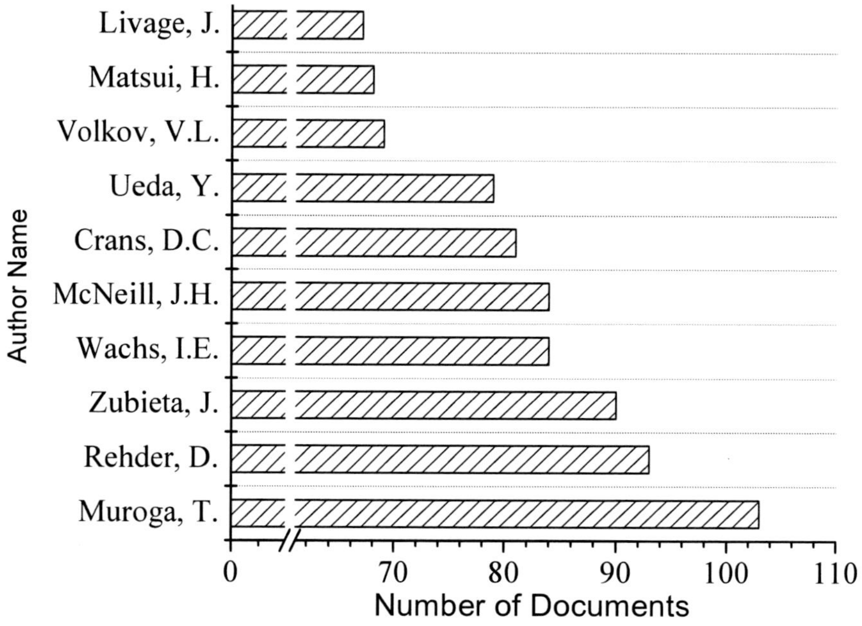

Figure 10. Ranking of the *most productive authors in* the field.

3.8. Keyword Frequency

Keyword frequency is the measure of the number of times a keyword appears on keywords list of publication. During the entire study period (Figure 11a), the most frequently used keyword was Vanadium with 14430 outputs between Paper and Review (40.54%) as it was also a keyword used in searching in Scopus database. Moreover, Vanadium Compounds (6662 outputs), Vanadium Alloys (3741 outputs), Chromium (2716 outputs), Titanium (2352 outputs), Nickel (2331 outputs), Aluminum (2184 outputs), Copper (2135 outputs), Iron (2069 outputs), Manganese (1996 documents), Zinc (1905 outputs), Molybdenum (1725 outputs), Cobalt (1570 outputs) are the most frequently used keywords.

This combination of keywords in the research output is easy to explain in terms of industrial applicability of today in making alloys and compounds for metallurgical applications. In fact, it is well known that one of the most important uses of vanadium as an alloying element in the steel industry and in the chemical industry as catalysts for a wide variety of products.

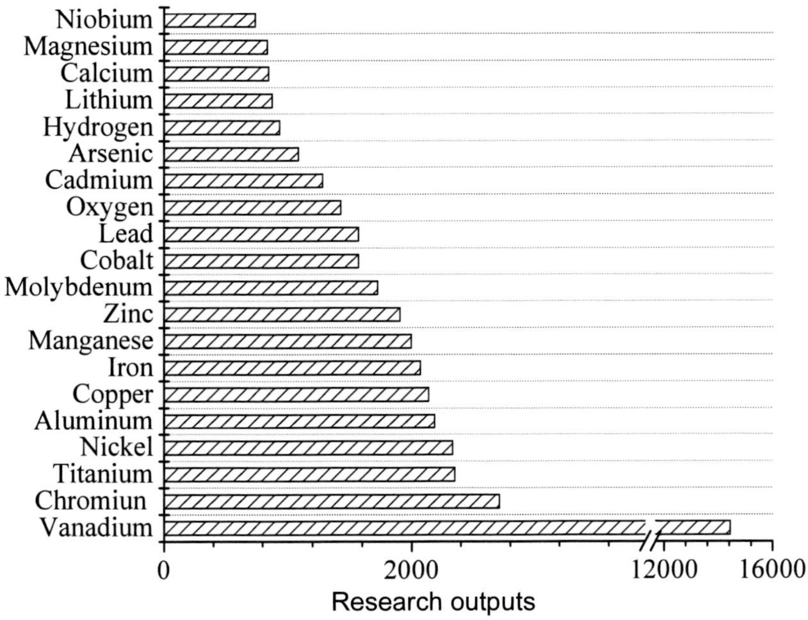

(a)

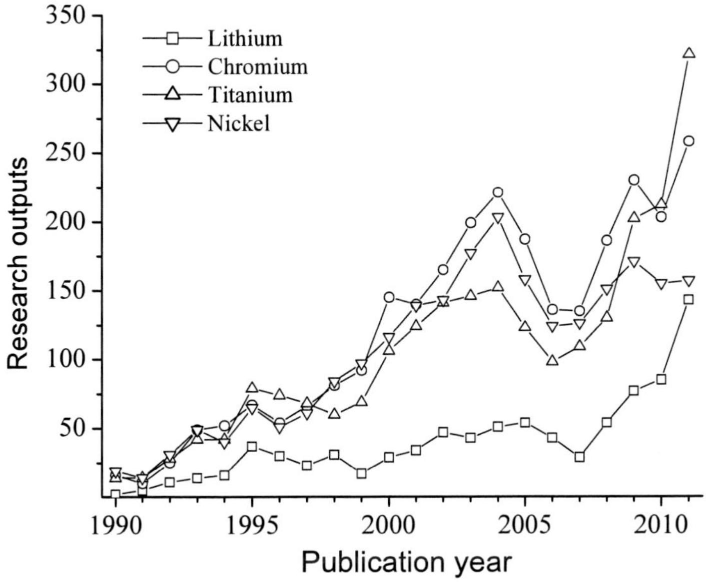

(b)

Figure 11. (a) Number of published research outputs as of the end of January 2012, disaggregated by chemical elements. (b) Growth in research output over time for selected chemical elements.

3.9. Citations Analysis

I selected the 22 most cited articles in each year published since 1990 for examination. The top most-cited articles in vanadium related research for the period 1990-2011 were tabulated in Table 1.

Table 1. Most cited research outputs (Article, Review and Conference Paper) in each year published since 1990

Year	Doc. type	Number of Authors	Title	Journal	Citations	Ref.
1990	Article	3	Laser raman spectroscopy of supported vanadium oxide catalysts	*Journal of Physical Chemistry* 94 (10), 4240-4246	248	[1]
1991	Review	1	Vanadium pentoxide gels	*Chemistry of Materials* 3 (4), 578-593	463	[2]
1992	Article	6	Physicochemical characterization of V-silicalite	*Journal of Physical Chemistry* 96 (6), 2617-2629	252	[3]
1993	Article	5	An inorganic double helix: Hydrothermal synthesis, structure, and magnetism of chiral $[(CH_3)_2NH_2]K_4[V_{10}O$ $10(H_2O)_2(OH)_4(PO_4)$ $7]\cdot 4H_2O$	*Science* 259 (5101), 1596-1599	265	[4]
1994	Article	11	Peroxovanadium compounds. A new class of potent phosphotyrosine phosphatase inhibitors which are insulin mimetics	*Journal of Biological Chemistry* 269 (6), 4596-4604	399	[5]
1995	Review	2	Oxidative dehydrogenation of lower alkanes on vanadium oxide-based catalysts. The present state of the art and outlooks	*Applied Catalysis A, General* 127 (1-2), 1-40	332	[6]
1996	Article	6	Giant voids in the hydrothermally synthesized microporous square pyramidal-tetrahedral framework vanadium phosphates $[HN(CH_2CH_2)$ $3NH]K_{1.35}[V_5O_9(PO_4)_2]\cdot xH_2O$ and $Cs_3[V_5O_9(PO_4)_2]\cdot xH_2O$	*Chemistry of Materials* 8 (1), 43-53	213	[7]
1997	Article	2	Oxidative dehydrogenation of short chain alkanes on supported vanadium oxide catalysts	*Applied Catalysis A: General* 157 (1-2), 117-142	432	[8]
1998	Article	6	Redox-active nanotubes of vanadium oxide	*Angewandte Chemie* 37 (9), 1263-1265.	285	[9]
1999	Article	3	Vanadium Compounds as Insulin Mimics	*Chemical Reviews* 99 (9), 2561-2571	429	10]
2000	Article	6	Inorganic-organic hybrid materials constructed from $[(VO_2)(HPO_4)]_\infty$ helical chains and $[M(4,4'\text{-bpy})_2]^{2+}$ (M = Co, Ni) fragments	*Angewandte Chemie* 39 (13), 2325-2327	213	[11]
2001	Article	2	Large remanent polarization of vanadium-doped $Bi_4Ti_3O_{12}$	*Applied Physics Letters* 78 (13), 1903-1905	303	[12]
2002	Conf. Paper	5	Antidiabetic vanadium(IV) and zinc(II) complexes	*Coordination Chemistry Reviews* 226 (1-2), 187-198	213	[13]
2003	Conf. Paper	2	Chemistry, spectroscopy and the role of supported vanadium oxides in heterogeneous catalysis	*Catalysis Today* 78 (1-4 SPEC.), 25-46	340	[14]

Year	Doc. type	Number of Authors	Title	Journal	Citations	Ref.
2004	Review	4	The chemistry and biochemistry of vanadium and the biological activities exerted by vanadium compounds	*Chemical Reviews* 104 (2), 849-902	427	[15]
2005	Article	4	Self-assembled vanadium pentoxide (V_2O_5) hollow microspheres from nanorods and their application in lithium-ion batteries	*Angewandte Chemie* 44 (28), 4391-4395	243	[16]
2006	Review	2	Synthesis and enhanced intercalation properties of nanostructured vanadium oxides	*Chemistry of Materials* 18 (12), 2787-2804	119	[17]
2007	Article	12	Mott transition in VO_2 revealed by infrared spectroscopy and nano-imaging	*Science* 318 (5857), 1750-1753	171	[18]
2008	Article	8	Experimental and theoretical study of the reactions between neutral vanadium oxide clusters and ethane, ethylene, and acetylene	*Journal of the American Chemical Society* 130 (6), 1932-1943	60	[19]
2009	Article	7	Transition of stoichiometric Sr_2VO_3 FeAs to a superconducting state at 37.2 K	*Physical Review B-Condensed Matter and Materials Physics* 79 (22), art. no. 220512	82	[20]
2010	Article	5	Nano-optical investigations of the metal-insulator phase behavior of individual VO_2 microcrystals	*Nano Letters* 10 (5), 1574-1581	26	[21]
2011	Article	7	High-performance supercapacitors based on intertwined CNT/V_2O_5 nanowire nanocomposites	*Advanced Materials* 23 (6), 791-795	11	[22]

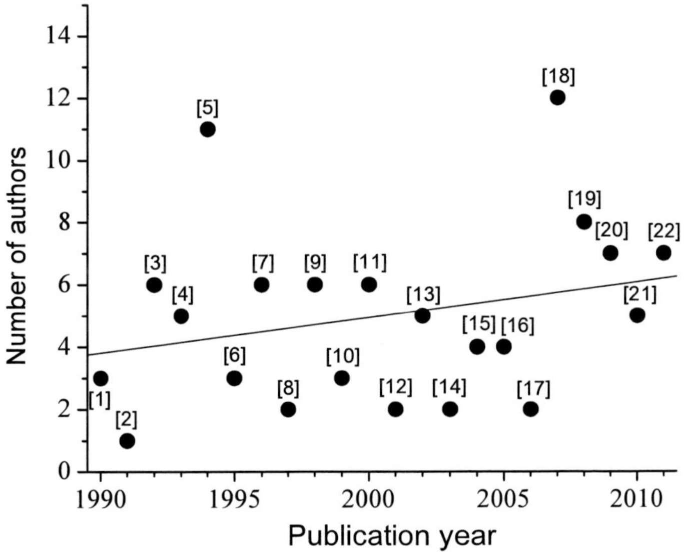

Figure 12. The number of co-authors of 22 most cited articles in each year published since 1990 (see Table 1).

Although caution is necessary to draw conclusions from the ranking data, these are useful for some interesting considerations. One of the most noticeable aspects of Table 1 is the distribution of the top cited articles among various fields of vanadium research [1-22]. All together, a total of 110 scholars authored the top 22 most cited articles (1990-2011) in the 14 journals. The *Science* was associated with the most authors per trial [18] and the *Chemistry of Materials* with the fewest [2].

A significant difference was noted between the numbers of authors associated with different document type; the median number of authors per Review in the vanadium was 2.5, compared with 3.5 in the Conference Papers. Article show an average of 5.8 authors per article. There was not relation between number of citations and authors. Moreover, *there was a weak positive correlation of* growth rates of number of authors per research outputs since 1990. One possible explanation for the trend towards increasing authorship includes the development of novel research strategies in vanadium subject by multidisciplinary teams, with subsequent proliferation of multi-authors per research outputs.

Conclusion

This chapter presents a review of the important literature related to vanadium in the last 20 years. It seems quite clear that - taking into account information from the Scopus database - the notable findings in the study include the fast increase of the number of research outputs on the subject of this book. It is equally clear that only five countries have contributed more than half of the total research outputs in the field of vanadium. Of course there are other ways to define scientific productivity but the scientometric study of the vanadium shows that The People's Republic of China is the major producer of scientific output from 2005 to today. Thus, it can be concluded that the research in this field is closely tied to economy growth. This is the context for the observations in this chapter.

Finally, looking to the future, the reader can play with probabilities. As an example, there is a high probability that the next research output is writing in English by a Chinese researcher focuses *on the Materials Science* aspects of vanadium alloy and it will contain more of 4 co-authors. *There is also nearly one chance* in five that Chromium and Titanium metals will stay in the list of keywords.

References

[1] Went, G.T.; Ted Oyama, S.; Bell, A.T. Laser raman spectroscopy of supported vanadium oxide catalysts. *Journal of Physical Chemistry* 1990, 94 (10), 4240-4246.

[2] Livage, J. Vanadium pentoxide gels. *Chemistry of Materials* 1991, 3 (4), 578-593.

[3] Centi, G.; Perathoner, S.; Trifiró, F.; Aboukais, A.; Aïssi, C.F.; Guelton, M. Physicochemical characterization of V-silicalite. *Journal of Physical Chemistry* 1992, 96 (6), 2617-2629.

[4] Soghomonian, V.; Chen, Q.; Haushalter, R.C.; Zubieta, J.; O'Connor; C.J. An inorganic double helix: hydrothermal synthesis, structure, and magnetism of chiral

$[(CH_3)_2NH_2]K_4[V_{10}O\ 10(H_2O)_2(OH)_4(PO_4)\ _7]\cdot 4H_2O$. *Science* 1993, 259 (5101), 1596-1599.
[5] Posner, B.I.; Faure, R.; Burgess, J.W.; Bevan, A.P.; Lachance, D.; Zhang-Sun, G.; Fantus, I.G.; Ng, J.B.; Hall, D.A.; Lum, B.S.; Shaver, A. Peroxovanadium compounds. A new class of potent phosphotyrosine phosphatase inhibitors which are insulin mimetics. *Journal of Biological Chemistry* 1994, 269 (6), 4596-4604.
[6] Mamedov, E.A.; Cortés Corberán, V. Oxidative dehydrogenation of lower alkanes on vanadium oxide-based catalysts. The present state of the art and outlooks. *Applied Catalysis A, General* 1995, 127 (1-2), 1-40.
[7] Khan, M.I.; Meyer, L.M.; Haushalter, R.C.; Schweitzer, A.L.; Zubieta, J.; Dye, J.L. Giant voids in the hydrothermally synthesized microporous square pyramidal-tetrahedral framework vanadium phosphates $[HN(CH_2CH_2)\ 3NH]K_{1.35}[V_5O_9(PO_4)_2]\cdot xH_2O$ and $Cs_3[V_5O_9(PO_4)\ _2]\cdot xH_2O$. *Chemistry of Materials* 1996, 8 (1), 43-53.
[8] Blasco, T.; López Nieto, J.M. Oxidative dehydrogenation of short chain alkanes on supported vanadium oxide catalysts. *Applied Catalysis A: General* 1997, 157 (1-2), 117-142.
[9] Spahr, M.E.; Bitterli, P.; Nesper, R.; Müller, M.; Krumeich, F.; Nissen, H.U. Redox-active nanotubes of vanadium oxide. *Angewandte Chemie - International Edition* 1998, 37 (9), 1263-1265.
[10] [Thompson, K.H.; McNeill, J.H.; Orvig, C. Vanadium Compounds as Insulin Mimics. *Chemical Reviews* 1999, 99 (9), 2561-2571.
[11] Shi, Z.; Feng, S.; Gao, S.; Zhang, L.; Yang, G.; Hua, J. Inorganic-organic hybrid materials constructed from $[(VO_2)(HPO_4)]_\infty$ helical chains and $[M(4,4'\text{-bpy})_2]^{2+}$ (M = Co, Ni) fragments. *Angewandte Chemie - International Edition* 2000, 39 (13), 2325-2327.
[12] Noguchi, Y.; Miyayama, M. Large permanent polarization of vanadium-doped $Bi_4Ti_3O_{12}$. *Applied Physics Letters* 2001, 78 (13), 1903-1905.
[13] Sakurai, H.; Kojima, Y.; Yoshikawa, Y.; Kawabe, K.; Yasui, H. Antidiabetic vanadium(IV) and zinc(II) complexes. *Coordination Chemistry Reviews* 2002, 226 (1-2), 187-198.
[14] Weckhuysen, B.M.; Keller, D.E. Chemistry, spectroscopy and the role of supported vanadium oxides in heterogeneous catalysis. *Catalysis Today* 2003, 78 (1-4), 25-46.
[15] Crans, D.C.; Smee, J.J.; Gaidamauskas, E.; Yang, L. The chemistry and biochemistry of vanadium and the biological activities exerted by vanadium compounds. *Chemical Reviews* 2004, 104 (2), 849-902.
[16] Cao, A.-M.; Hu, J.-S.; Liang, H.-P.; Wan, L.-J. Self-assembled vanadium pentoxide (V_2O_5) hollow microspheres from nanorods and their application in lithium-ion batteries. *Angewandte Chemie - International Edition* 2005, 44 (28), 4391-4395.
[17] Wang, Y.; Cao, G. Synthesis and enhanced intercalation properties of nanostructured vanadium oxides. *Chemistry of Materials* 2006, 18 (12), 2787-2804.
[18] Qazilbash, M.M.; Brehm, M.; Chae, B.-G.; Ho, P.-C.; Andreev, G.O.; Kim, B.-J.; Yun, S.J.; Balatsky, A.V.; Maple, M.B.; Keilmann, F.; Kim, H.-T.; Basov, D.N. Mott transition in VO_2 revealed by infrared spectroscopy and nano-imaging. *Science* 2007, 318 (5857), 1750-1753.

[19] Dong, F.; Heinbuch, S.; Xie, Y.; Rocca, J.J.; Bernstein, E.R.; Wang, Z.-C.; Deng, K.; He, S.-G. Experimental and theoretical study of the reactions between neutral vanadium oxide clusters and ethane, ethylene, and acetylene. *Journal of the American Chemical Society* 2008, 130 (6), 1932-1943.

[20] Zhu, X.; Han, F.; Mu, G.; Cheng, P.; Shen, B.; Zeng, B.; Wen, H.-H. Transition of stoichiometric Sr_2VO_3FeAs to a superconducting state at 37.2 K. *Physical Review B - Condensed Matter and Materials Physics* 2009, 79 (22), No. 220512.

[21] Jones, A.C.; Berweger, S.; Wei, J.; Cobden, D.; Raschke, M.B. Nano-optical investigations of the metal-insulator phase behavior of individual VO_2 microcrystals. *Nano Letters* 2010, 10 (5), 1574-1581.

[22] Chen, Z.; Augustyn, V.; Wen, J.; Zhang, Y.; Shen, M.; Dunn, B.; Lu, Y. High-performance supercapacitors based on intertwined CNT/V_2O_5 nanowire nanocomposites. *Advanced Materials* 2011, 23 (6), 791-795.

[23] Jeon S.I.; Park J.H.; Magnone E.; Lee Y.T.; Fleury E. Hydrogen permeation of Pd-coated $V_{90}Al_{10}$ alloy membranes at different pressures in the presence and absence of carbon dioxide. *Current Applied Physics* 2012, 12 (2), 394-400.

[24] Magnone E.; Jeon, S.I.; Park J.H.; Fleury E.. Hydrogen permeation properties and chemical stability of novel Pd-free alloy membranes based on the V-Y system. *Chemical Engineering & Technology* 2011, Accepted, In press.

[25] Tague-Sutcliffe J. An introduction to informetrics. *Information Processing & Management* 1992. 28(1), 1-3.

[26] Magnone E. An analysis for estimating the short-term effects of Japan's triple disaster on progress in materials science. *Journal of Informetrics* 2012, Accepted, In press.

[27] Bar-Ilan J. Informetrics at the beginning of the 21st century - a review. *Journal of Informetrics* 2008, 2, 1-52.

[28] Scopus. SciVerse Scopus, Elsevier B.V. *http://www.scopus.com/home.url*

[29] Scopus, SciVerse Scopus, Elsevier B.V. (downloaded on 15 January 2012 from *http://www.info.sciverse.com/UserFiles/sciverse_scopus_content_coverage_0.pdf)*

[30] Vieira, E.S.; Gomes, J.A.N.F. A comparison of Scopus and Web of Science for a typical university. *Scientometrics* 2009, 81, 587-600.

[31] Jacso, P. As we may search – comparison of major features of the Web of Science, Scopus, and Google Scholar citation-based and citation-enhanced databases. *Current Science* 2005, 89, 1537-1547.

[32] Zhou, P.; Leydesdorff, L. The emergence of China as a leading nation in science. *Research Policy* 2006, 35 (1), 83-104.

[33] The Royal Society (2011). Knowledge, networks and nations - Global scientific collaboration in the 21st century. RS Policy document 03/11, Issued: March 2011 DES2096, London, UK (downloaded on 15 January 2012 from *http://royalsociety.org/uploadedFiles/Royal_Society_Content/Influencing_Policy/Reports/2011-03-28-Knowledge-networks-nations.pdf)*

[34] Adams J.; King C.; Ma N. *Global research report China - Research and collaboration in the new geography of science.* Evidence Ltd, Thomson Reuters company (November 2009) Leeds, UK.

[35] Leydesdorff L.; Zhou P. Are the contributions of China and Korea upsetting the world system of science? *Scientometrics* 2005, 63 (3) 617-630.

[36] Muroga, T.; Nagasaka, T.; Abe, K.; Chernov, V.M.; Matsui, H.; Smith, D.L.; Xu, Z.-Y.; Zinkle, S.J. Vanadium alloys - overview and recent results. *Journal of Nuclear Materials* 2002, 307-311, 547-554.

[37] Rehder, D. The bioorganic chemistry of vanadium. *Angewandte Chemie - International Edition* 1991, 30 (2), 148-167.

In: Vanadium
Editors: V. N. Baranova and A. V. Fortunatov

ISBN 978-1-62081-600-4

Chapter 6

VANADIUM

Maria Rosa Avila-Costa
Department of Neuroscience, Neuromorphology Lab
UNAM, Mexico

1. GENERAL DESCRIPTION

Vanadium (V) is a transition element, is a member of Group 5, and is the 23rd element of the Periodic table. Andres Manuel Del Rio was the first chemist to provide the idea of this new element in 1801. But it was discovered by Nils Sefstrom, a Swedish chemist in 1830 (Mukherjee et al. 2004). Pure V is a bright sliver-white, soft and ductile metal and 22nd most abundant element in the earth's crust. V has become subject of interest amongst nutritionists since the discovery that various marine species have this metal as an essential element (Almedeida et al. 2001). Although most food contains low amount of V (<1 ng/g), food is the major source of exposure to V for general population (Barceloux, 1999).

V may be beneficial and possibly essential in humans, but certainly essential for some living organisms (Kustin et al. 1983; Michibata and Sakurai, 1990; Nielsen and Uthus, 1990; Rehder, 1999; Michibata et al. 2003a,b; Crans et al. 2004). The usual oxidation states of the vanadium are V (III), V (IV) and V (V). The V ions play a role in biology as counter ions for various proteins, DNA and RNA.

V is widely distributed on Earth, however its role as a micronutrient in humans is not yet established. Humans are exposed to V mainly through the polluted atmosphere from combustion products of vanadium-bearing fuel oils, fumes and dust. Food contains a very low content of V usually below 1 ng/g. V enters the organism by inhalation, skin and gastrointestinal tract and accumulates mainly in the liver, kidney, spleen, bones and to a lesser extent in lungs and brain (Hansen et al. 1982; Bracken et al. 1985; Barceloux, 1999).

The structural role is often manifested by the maintenance of various biological structures, whereas a functional role is to bring key reactivity to a reaction center for a protein (Crans et al. 2004). V ions have many structural roles reflected by its structural and electronic analogy to phosphorus (Rehder, 1999). In addition, the V ion is an enzyme cofactor, (Van Pee et al. 2000; Rehder, 2003) and is found in certain tunicates (Michibata et al. 2003b) and possibly mammals (Nielsen and Uthus, 1990). During the last few decades, the facade of V as a 'slightly' toxic and carcinogenic element eventually ratified to an essential trace element with anti−diabetic and anti-carcinogenic properties (Mukherjee et al. 2004).

Neurotoxic effects of V are not well understood yet, but it is known that acute poisoning in animals by ingestion or inhalation of V compounds leads to nervous disturbances, paralysis of legs, respiratory failure, convulsions, bloody diarrhea, and death (Wenning and Kirsch, 1988). V penetrates the blood–brain barrier (Berman, 1980; Avila-Costa et al. 2005) and induces neurochemical alterations: noradrenaline, dopamine, and 5-hydroxytriptamine altered levels and an inhibitory effect on the uptake and release of noradrenaline were found in the rat brain during V poisoning (Witkowska and Brzezinski, 1979; 1983; Avila-Costa et al. 2004).

It is well established that vanadate (V^{5+}) and vanadyl (V^{4+}), the V oxidation states of biological interest, may cause a number of adverse toxic effects in mammals depending on circulating V levels. Industrial exposure of metal workers to vanadium pentoxide (V_2O_5) leads to cardiovascular diseases and a variety of symptoms involving central nervous system (CNS), gastrointestinal and respiratory systems (WHO, 1988). Moreover, it has been suggested that raised tissue levels of V may be of etiological importance in manic-depressive illness (Nechay, 1984; Naylor et al. 1984; Conri et al. 1986; Campbell et al. 1988). Serum V levels in adults with depression were significantly greater than those of controls, regardless of the type of depression, tremor and impaired conditioned reflexes, as well as congestion of brain spinal cord (WHO, 1988). Moreover, reduced cognitive abilities in humans chronically exposed to this element were found (Barth et al. 2002).

2. Sources

Metallic V does not occur in nature. Over 70 V minerals are known, carnatite and vanadinite being the most important from the point of view of mining. V is also found in phosphate rock and certain iron ores, and is present in some crude oils in the form of organic complexes. It is also found in small percentages in meteorites. Production of V is linked with that of other metals such as iron, uranium, titanium, and aluminum. As rich minerals rarely occur in large deposits, ores with low V content, which exist in large amounts, are important. Extraction of V from fossil fuels, including vanadium-rich oil and coal, tars, bitumens, and asphaltites is important in several countries (WHO, 1988). During the first half of the 1980s, the global production of V (as V_2O_5) ranged from 34 to 45 million Kg, China, Finland, South Africa, the USA, and the USSR being the biggest producers (WHO, 1988).

V is mainly (75 - 85%) used in ferrous metallurgy as an alloy additive in various types of steel. Its use in nonferrous metals and is important for the atomic energy industry, aircraft construction, and space technology. V is also widely used as a catalyst in the chemical industry, where V_2O_5 and metavanadates are especially important for the production of sulfuric acid and plastics. Small quantities of V are used in a variety of other applications. Emissions of V may be high in the vicinity of large plants producing steel alloys. V is also released into the air: during the re-smelting of scrap steel and the transformation of titaniferrous and vanadic magnetite iron ores into steel; from the roasting of V slags; from V_2O_5 smelting furnaces; and from electric furnaces in which ferrovanadium is smelted (WHO, 1988).

The global recycling of V involves the release of V from natural and anthropogenic sources to the air, water, and soil, the transport of particles of V in water and air, wet and dry deposition, absorption, and complexing (ATSDR, 1997).

2.1. Air

Exposure of the general population to V in air results primarily from the combustion of petroleum, coal, and heavy oils during the generation of electricity and heat. During the burning of oil fields in Kuwait, air particulate matter in Bahrain contained 11–42 ng V/m^3 (Madany and Raveendran, 1992).

Concentrations in coal, crude oil, and carbon fossils average approximately 100 μg V/g, 600 μg V/g, and 1500 μg V/g, respectively (Meish and Bielig, 1980). Very little V is present in natural gas. Additional industrial sources of V in the metallurgy industry include discharges from furnaces roasting V slag, V_2O_5 smelting furnaces, and crucibles, which melt ferrovanadium alloys. Anthropogenic sources account for about two thirds of the atmospheric V, which occurs in the form of V oxides (ATSDR, 1997). Continental dust, marine aerosols, and, to a lesser extent, volcanic emissions account for the rest of the V in the atmosphere. Vanadium enters the air as an aerosol of simple or complex V oxides from anthropogenic sources and as mineral particles of less soluble trivalent forms from natural sources (Barceloux, 1999).

2.2. Soil

The actual content of V depends on the type of parent rock. The soil types, which contain the highest concentration of V, are Tundra podzols and clays. Elemental V does not occur in nature because of its propensity to react with other elements, particularly oxygen (ATSDR, 1997). The largest releases of V to the soil result from the weathering of rocks rather than from anthropogenic sources, such as super-phosphate fertilizers (0.05–2 g V/kg) and basic slag (1–5 g V/kg) (Byerrum et al. 1974).

Basic rocks contain higher concentrations of V compared with neutral and acidic rocks.

2.3. Water

Drinking water is not an important source of exposure to V for the general population with typical V concentrations <1 μg/L drinking water (WHO, 1988). Natural releases of V from the erosion of soil and the weathering of rock far exceed the deposition of V from anthropogenic sources in the air. Most samples from drinking water supplies contain levels of V below detectable limits. When detected, the concentration in drinking water usually is 1–6 μg V/L with a range up to about 20 μg V/L (WHO, 1988). The level of V in freshwater depends on geographical difference in leachates and effluents from both natural and anthropogenic sources (Barceloux, 1999).

2.4. Food

Even though most foods contain low concentrations (<1 ng V/g) (Byrne and Kosta, 1978), food is the major source of exposure to V for the general population. The daily requirement for V probably is small (<10 μg/d) with the highest source of V present in black pepper, dill seed, mushrooms (0.05–2 μg/g), parsley (1.8 μg/g), shellfish, spinach (0.5–0.8 μg/g), and some prepared foods (Nielsen, 1991).

In general, seafood contains higher concentrations of V compared with terrestrial animal sources. Beverages, fresh fruit and vegetables, cereals, liver, fats, and oil contain smaller amounts (<1–10 ng V/g). The processing of food tends to raise the concentrations of V (Myron et al. 1977). High concentrations of V exist in tobacco, and smoke from tobacco contains 1–8 ppm V (Byrne and Kosta, 1978). The mushroom, *Amanita muscaria* contains about 100 times the concentration (100 ppm V) found in other plants and mushrooms (Bertrand, 1950). The estimated daily dietary intake for the US population is about 10–60 μg V (Harland and Harden-Williams, 1994).

2.5. Occupational Exposure

V levels near metallurgical industries usually average about 1 mg V/m^3 whereas ambient air near industries, which produce V metal or compounds, contains concentrations of V of a few mg V/m^3 (WHO, 1988). Very high levels of V result from boiler-cleaning operations as a result of the high concentration (approximately 10–25%) of V oxides in the dust. During these cleaning operations, concentrations of 50–100 mg V/m^3 are frequent with concentrations ranging up to 500 mg V/m^3 (Barceloux, 1999).

In terms of occupational exposure, the most important V compounds are V_2O_5, vanadium trioxide, ferrovanadium, vanadium carbide, and vanadium salts, such as sodium and ammonium vanadate. The oxides and salts are commonly used in industry in powder form, giving rise to the possibility of dust and aerosol formation, when the substances are crushed or ground. Many metallurgical processes involve the production of vapor containing V_2O_5, which condenses to form respirable aerosols.

Boiler-cleaning operations generate dusts containing the pentoxide and trioxide compounds. Combustion of residual fuels with high V content is likely to produce aerosols of the pentoxide as well as oxide complexes of V with other metals (WHO, 1988).

3. Absorption, Distribution and Excretion

Absorption, excretion and storage mechanisms of V in living system are not thoroughly understood. Several reports show that V is poorly (only about 10%) absorbed from gastrointestinal tract (Nriagu, 1998; Poucheret et al. 1998). These reports suggests that most of the ingested V is transformed into the cationic vanadyl form in stomach before being absorbed in the duodenum through an unidentified mechanism (Hirano and Suzuki, 1996). Again, V in its anionic vanadate form has been found to be absorbed at much higher quantities (about five times more than vanadyl form) through anionic transport system

(Hirano and Suzuki, 1996). It has also been reported that vanadyl undergoes spontaneous oxidation to vanadate *in vivo* (Li et al. 1996). Multivalent existence of V in nature and in living systems put forth the chemical complexity of this element. This multifaceted chemical character of V in turn echoes in its biological and biochemical properties, especially in metabolism and in absorption. Again vanadate after reaching the blood stream is converted into vanadyl ion, although the vanadate form also exists. Thus, vanadate (by transferrin) and vanadyl (by albumin and transferrin) are rapidly transported by blood proteins to various tissues (Fantus et al. 1995). Blood parameters showed little or no reflection of toxicity after a long-term supplementation of V compounds (Fawcett et al. 1997; Guidotti et al. 1997), which might be due to the brisk transport of V from blood to the tissues. Upon supplementation, V is incorporated in various organs and tissues including liver, kidney, brain, heart, muscles and bone. Kidney, spleen, bone and liver tissues of rat have been shown to accumulate distinctly high amounts of V in chronically treated animals through oral administration (Hamel and Duckworth, 1995; Ramandham et al. 1991).

The effects of V administration persist even after V has been withdrawn for several days (Cam et al. 2000). Unabsorbed V is excreted in feces. When V was administrated through parenteral route, 10% of the V was found in the feces of humans and rats (Barceloux, 1999; Setyawati et al. 1998; Alimonti et al. 2000). V is excreted through bile and through urine (Alimonti et al. 2000). It is, thus, the bile route through which a significant amount of V may be ultimately excreted through feces. Moreover, it may be suggested that V content in feces does not reflect the amounts of V absorbed or unabsorbed (Mukherjee et al. 2004). Figure 1 summarizes the kinetics of Vanadium.

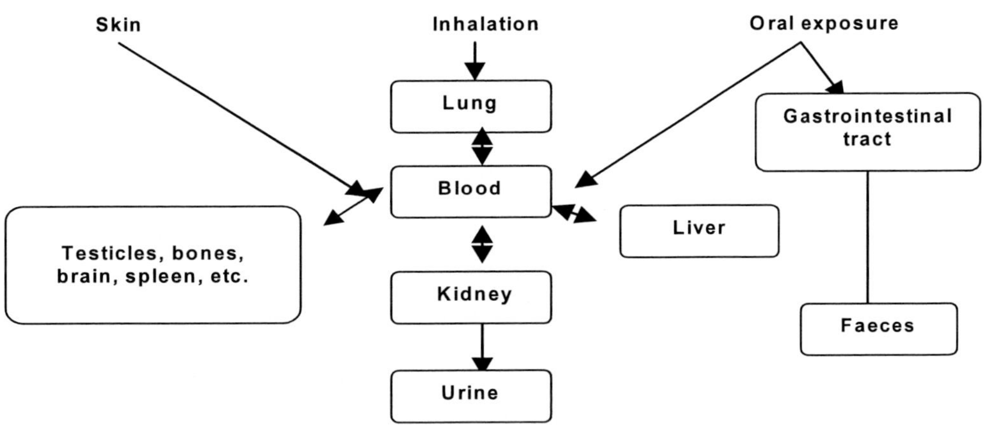

Figure 1. Vanadium kinetics.

4. General Toxicity

The toxicity of V compounds depends on a variety of factors including the route of administration and the inherent toxicity of the particular compound. In general, the toxicity of V compounds is low, and the toxicity is least following ingestion and greatest following parenteral administration. Inhalation is a route of exposure that produces intermediate toxicity

(WHO, 1988; Barceloux, 1999; National Toxicology Program, 2002). The toxicity of V increases with higher valences and the pentavalent compounds are usually the most toxic. Acidification tends to reduce the toxicity of V compounds (Mitchell, 1953).

Studies in animals have shown that equivalent doses of V_2O_5 are better tolerated by small animals, including rats and mice, than by larger animals, such as rabbits and horses (Hudson, 1964). The LD50 of V_2O_5 is highly species-dependent. Differences in diet and route of V administration may contribute to these discrepancies (IARC, 2006).

In humans, acute V poisoning can manifest itself in a number of symptoms including eye irritation and tremors of the hands (Lewis, 1959). In addition, a greenish coloration of the tongue has been observed in humans exposed to high concentrations of V_2O_5 and is probably due to the formation of trivalent and tetravalent V complexes (Wyers, 1946). The green color disappears within 2–3 days of cessation of exposure (Lewis, 1959).

4.1. Mechanism of Action

The systemic effects of V probably result from its oxidizing ability (Kiviluoto et al. 1981). V also inhibits oxidative phosphorylation (Hathcock et al. 1966), but the role of this mechanism in toxicity is unclear. Vanadate interferes with phosphate-containing enzymes (ATP phosphohydrolases, adenylate kinase, glyceraldehyde-3-phosphate dehydrogenase, ribonuclease) at relatively high (μmol-mmol) extracellular concentrations (Cantley et al. 1977). Based on *in vitro* data, vanadate probably is a noncompetitive inhibitor of Na K-ATPase (K1= 880 nM). This species of V is a more effective inhibitor of ATPases in concentrations >50 μM, whereas vanadyl is an effective inhibitor in very low concentrations (10 nM) (Janiszewska et al. 1994).

5. Nervous System Effects

Neurophysiological effects have been reported following acute exposure (by oral administration and subcutaneous injection) of dogs and rabbits to V oxides and salts (vanadium trioxide, V_2O_5, vanadium trichloride and ammonium metavanadate) (IARC, 2006). These effects included disturbances of the CNS, such as impaired conditioned reflexes and neuromuscular excitability (Roshchin, 1967). The animals behaved passively, refusing to eat, and lost weight. In cases of severe poisoning, diarrhea, paralysis of the hind limbs and respiratory failure were followed by death (Roshchin, 1967).

In a study reported by Seljankina (1961 cited by WHO, 1988), solutions of V_2O_5 were administered orally to rats and mice at doses of 0.005–1 mg/Kg per day for periods ranging from 21 days at the higher concentrations to 6 months at the lower concentrations. A dose of 0.05 mg/Kg was found to be the threshold dose for functional disturbances in conditioned reflex activity in both mice and rats. Repeated exposure to aqueous solutions (0.05–0.5 mg/Kg per day, for 80 days) of V_2O_5 impaired conditioned reflex mechanisms in rats.

In male CD-1 mice exposed by inhalation to 0.02 M V_2O_5 2h twice a week for 4 weeks, Golgi staining revealed a drastic reduction in dendritic spines in the striatum compared with controls, showing that the inhalation of V_2O_5 causes severe neuronal damage in the corpus striatum (Avila-Costa et al. 2004). Using the same inhalation model, after 12 weeks of exposure, a decrease in dendritic spines of granule cells of the olfactory bulb was observed (Mondragón et al. 2003). In addition, ultrastructural modifications in nuclear morphology of these cells were evident, Golgi apparatus was dilated and an increase in lipofucsin granules was observed, as well as necrosis of some cells (Colin-Barenque et al. 2003). In the cerebellum, necrosis and apoptosis of the Purkinje and granule cell layers were seen (Meza et al. 2003). There is scant information about V and neurodegeneration, for that reason our group have developed an inhalation model in mice to evaluate different brain areas, starting with the olfactory bulb, which is one of the main entrances to the brain, followed by the cerebral ventricles and the consequences in striatum, substantia nigra and hippocampus, structures highly related with neurodegenerative disorders.

5.1. Olfactory Bulb

The ability to smell, like other primary sensory abilities, diminishes with age. Olfactory decrement is especially prominent in the elderly and in neurodegenerative diseases. This decrement is correlated with neuronal and synaptic loss in the olfactory bulb, which in some way correlates with exposure to airborne toxins (Mesholam et al. 1998; Ponsen et al. 2004; Jankovic, 2008). In this way, after V inhalation, we found that mice show evident olfactory bulb alterations, which include: reduction of dendritic spines density of granule cells (Figure 2 A and B) and apoptotic-like cell death (Figure 3).

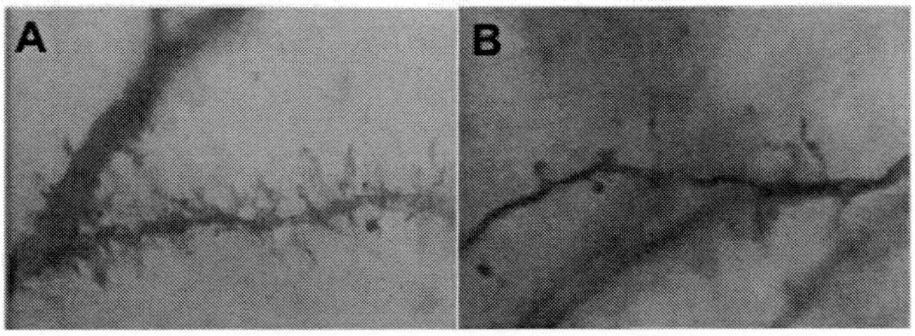

Figure 2. A. Control granule cell dendrite of the olfactory bulb in which is notorious the presence of spines. B. Evidences the dendritic spine loss after V inhalation. 4000 X.

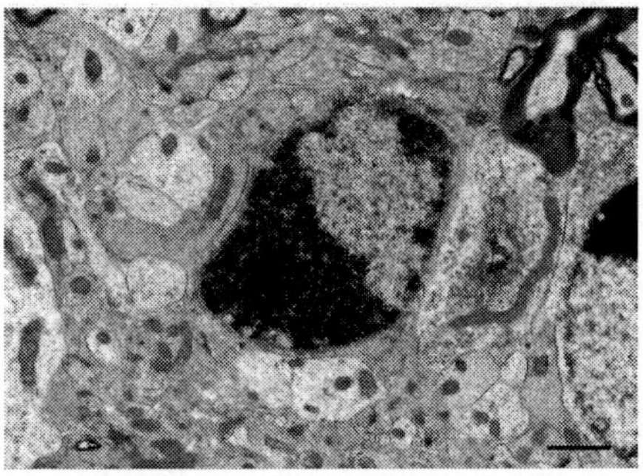

Figure 3. Apoptotic granule cell of exposed mouse evidenced by condensation and margination of the chromatin. Scale bar = 1.5 μm.

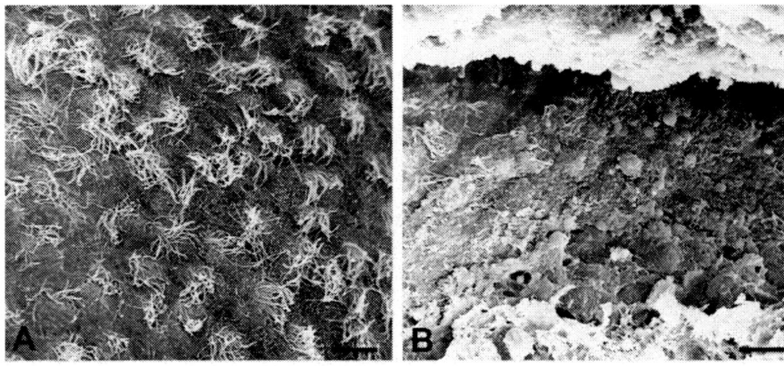

Figure 4. Scanning electron microscopy micrographs of the floor of the fourth ventricle. A. Control mouse and B. Vanadium exposed mouse; the micrograph evidences cilia loss. Bar 10µm.

5.2. Cerebral Ventricles

The blood–brain barrier (BBB) plays a pivotal role in the removal of endogenous and exogenous toxins. Regulation of blood–brain tissue exchange is accomplished by ependymal cells, which possess intercellular tight junctions. Loss of BBB function is an etiologic component of many neurological disorders. Cumulative evidence has revealed that the brain barriers are subject to toxic insults from heavy metal exposure (Zhen et al. 2002).

In our laboratory we investigate the effects of V_2O_5 inhalation on mouse fourth ventricle ependymal epithelium (Avila-Costa et al. 2005). Scanning electron microscopy analysis demonstrates loss of cilia from the area around the floor of the fourth ventricle, a tendency for the remaining cilia to conglomerate, sloughed cells and disintegrated structure compared to that of control mice (Figure 4 A and B); and transmission electron microscopy showed ependymal cell layer detachment from the basal membrane and dissolution of the tight junctions between ependymal cells (Figure 5 A and B).

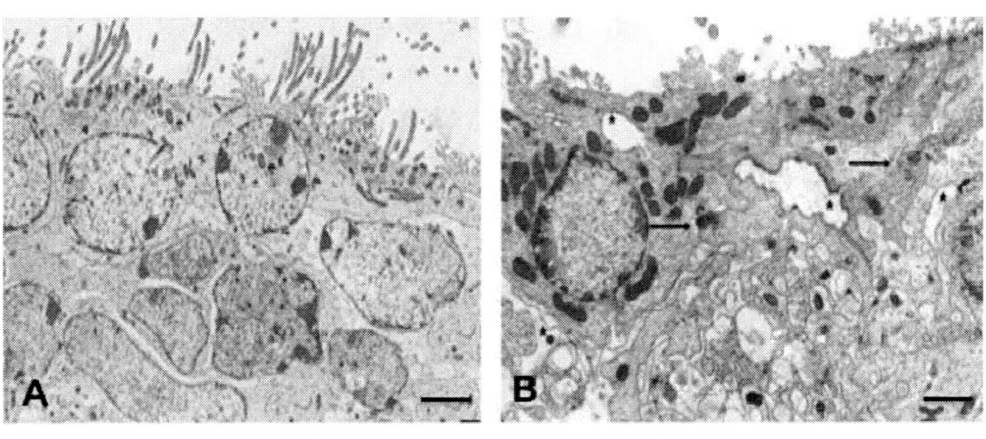

Figure 5. Transmission electron microscopy micrographs of control mouse (A) and V exposed mouse (B). Panel B shows the opening of the intracellular junctions (arrows), and ependymal cells detachment (asterisks) (ruthenium red). Bar 10µm.

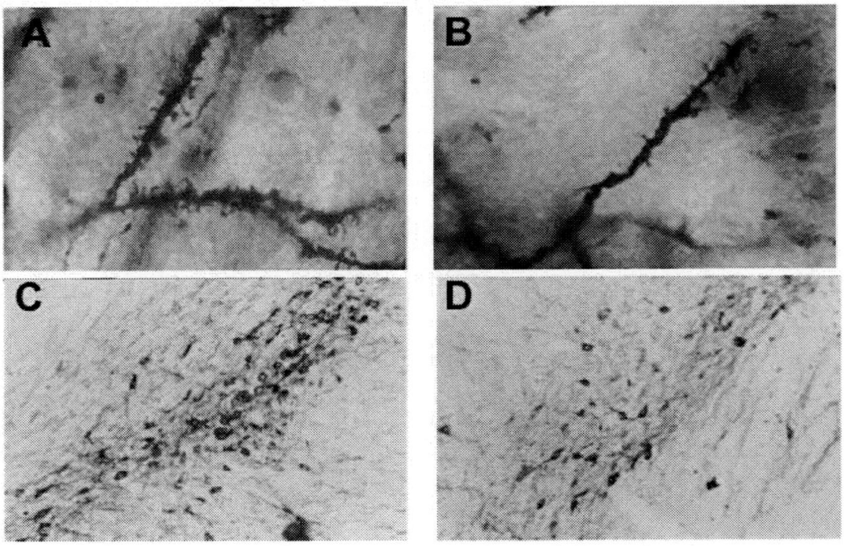

Figure 6. A. Control striatum medium size spiny neuron in which is notorious the presence of spines on the dendrites. B. Evidences the dendritic spine loss after V inhalation. C and D. Representative TH-immunostained coronal sections of substantia nigra of (C) control mouse and (D) exposed mouse. The cell loss due to the V inhalation was significantly different from control mice. 4000 X.

We conclude that this damage can allow toxicants to modify the permeability of the epithelium and promote access of inflammatory mediators to the underlying neuronal tissue causing injury and neuronal death. Since Berman (1980) establish that vanadate anion penetrates BBB and leads to CNS dysfunctions.

5.3. Basal Ganglia

Using the same model, we demonstrated that mice, which inhaled V_2O_5 had a significant dopaminergic neuronal loss in the substantia nigra and as a consequence, developed morphological alterations of the striatum medium size spiny neurons (Figure 6) (Avila-Costa et al. 2004). The relevance of our basal ganglia findings resides in the facts that demonstrate that V produces free radicals (Sasi et al. 1994) and interferes with the catecholaminergic brain levels (Sharma et al. 1986). As it is well established, the basal ganglia are brain structures especially vulnerable to oxidative stress as a consequence of its lower antioxidants concentrations, and the elevated presence of iron in the substantia nigra. The dopamine catabolism implies the generation of a great amount of free radicals, which in turn produces an oxidative stress state (Luo et al. 1998). Furthermore, there is evidence that oxidative stress is implicated with neurodegenerative diseases such as Parkinson (Simonian and Coyle, 1996).

5.4. Hippocampus

Recent findings from our group, demonstrate that mice which inhaled V, also manifest spatial memory deterioration as a consequence of hippocampal neuronal death (Avila-Costa et al. 2006). Results showed that V inhalation produces a time dependent loss of dendritic spines (Figure 7), necrotic-like cell death (Figure 8), and notorious alterations of the hippocampus CA1 neuropile, which correlate with spatial memory impairment (data not shown).

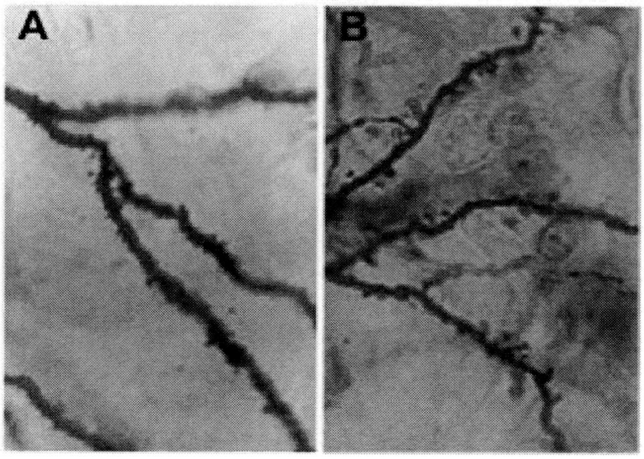

Figure 7. A control pyramidal cell of hippocampus CA1, it is notorious the presence of spines on the dendrites. B evidences the dendritic spine loss after V inhalation. 1000 X.

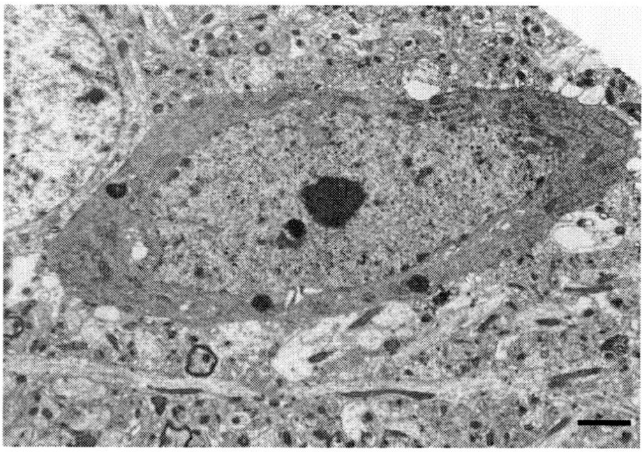

Figure 8. Necrotic pyramidal cell of exposed mouse, displaying electron-dense cytoplasm, dilated Golgi apparatus and endoplasmic reticulum. Scale bar = 1 μm.

It has been proposed that temporary inactivation or lesions of the dorsal hippocampus cause impairments in the acquisition and retrieval of spatial memory in tasks such as those evaluated in the Morris water maze. Substantial bodies of evidences indicate that hippocampal neuron loss is widely viewed as a hallmark of normal aging and to contribute directly to age-related deficits in learning and memory (Grady et al. 1995).

On the other hand, it has been agreed that many conditions lead to a decreased number of dendritic spines (Fiala et al. 2002). These conditions also demonstrate decreased neuronal number, suggesting that spines are lost as a result of a severe decline in the number and availability of axonal inputs to dendritic spines (Fiala et al. 2002). According with our results we assume, that the decreased number of dendritic spines and the number of necrotic cells are the attainable explanation for the memory impairments identified after V inhalation, since behavioral deficiency highly correlated with those hippocampal cell changes.

Furthermore, memory impairments is related with the phosphorylation of some signaling molecules, which have been proposed as one of the most prevalent mechanisms for modulating neuronal functions, including important mechanisms of memory acquisition (Kandel, 2001). It appears that V activates extracellular signal–regulated kinases (ERKs) trough competitive inhibition of protein tyrosine phosphatase (Zhao et al. 1996) that acts by irreversible oxidation of the catalytic site of these enzymes (Huyer et al. 1997), whose oxidation results in their inhibition (Augereau et al. 2005) affecting the tyrosine phosphorylation leading to cell alterations. Moreover, protein phosphorilation is involved in the regulation of neurotransmitter receptors such as NMDA, which mediates whole cell currents and intracellular Ca^{2+} and plays an essential role in memory mechanisms (Tlan Wang and Salter, 1994).

Final Considerations

The facts mentioned above provide evidences that V interferes with important cerebral mechanisms that are directly involved with the pathogenesis of neurodegenerative disorders. First, the olfactory bulb alterations which are considered as the first symptom of neurodegeneration, and also serves as the main entrance to the brain; second, the ependymal epithelium disruption, that allows toxicants to modify the permeability of the epithelium and promote access of inflammatory mediators to the underlying neuronal tissue causing injury and neuronal death; third, the fact that V interferes with dopaminergic brain concentrations leading to cell alterations in those structures related with Parkinson disease; and finally that V inhibits important enzymes which modulate neuronal functions, that leads to changes that include the generation of ROS which in turn produces cell death, and as a consequence functional alterations, such as memory deterioration, which is related to Alzheimer disease. Those findings would allow us planning strategies to protect the brain from toxicants such as V, which have increased in the atmosphere during the last decades and constitute an important health problem since metal pollution has been related with neurodegenerative diseases.

References

Alimonti, A., Petrucci, F., Krachler, M., Bocca, B. and Caroll, S. (2000) Reference values for chromium, nickel and vanadium in urine of youngsters form the urban area of Rome. *J. Environ. Monit.* 2: 351–354.

Almedeida, M., Filipe, S., Hunianes, M., Mala, M.F., Melo, R., Severin, N., Silva, J.A., Frueesto da Silva, J.J. and Wever, R. (2001) Vanadium haloperoxidases from brown algae of the Laminariaceae family. *Phytochemistry*, 57: 633–642.

ATSDR Agency for Toxic Substances and Disease Registry. (1997) ATSDR's Toxicological Profiles: Vanadium. Boca Raton, Florida: Lewis Publishers, CRC Press, Inc.

Avila-Costa, M.R., Colín-Barenque, L., Zepeda-Rodríguez, A., Antuna, S., Pasos, F., Saldivar, O.L., Espejel-Maya, G., Mussali-Galante, P., Avila-Casado, M del C. and Fortoul, T.I. (2005) Ependymal epithelium disruption after vanadium pentoxide inhalation: A mice experimental model. *Neurosci. Lett.* 381: 21–25.

Avila-Costa, M.R., Fortoul, T.I., Niño-Cabrera, G., Colín-Barenque, L., Bizarro-Nevares, P., Gutiérrez-Valdez, A.L., Ordóñez-Librado, J.L., Rodríguez-Lara, V., Mussali-Galante, P., Díaz-Bech, P. and Anaya-Martínez, V. (2006) Hippocampal cell alterations induced by the inhalation of vanadium pentoxide (V_2O_5) promote memory deterioration. *Neurotoxicolgy*, 27: 1007-1012.

Avila-Costa, M.R., Montiel Flores, E., Colin-Barenque, L., Ordóñez, J.L., Gutiérrez, A.L., Niño-Cabrera, H.G., Mussali-Galante, P. and Fortoul, T.I. (2004). Nigrostriatal modifications After Vanadium Inhalation: An Immunocytochemical and Cytological Approach. *Neurochem. Res.* 29: 1365–1369.

Augereau, O., Claverol, S., Boudes, N., Basurko, M.J., Bonneu, M., Rossignol, R., Mazat, J.P., Letellier, T. and Dachary-Prigent, J. (2005) Identification of tyrosine-phosphorylated proteins of the mitochondrial oxidative phosphorylation machinery. *Cel. and Mol. Life Sci.* 62: 1-11.

Barceloux, D.G. (1999) Vanadium. *J. Toxicol. Clin. Toxicol.* 37:265-78.

Barth, A., Schaffer, A.W., Konnaris, C., Blauensteiner, R., Winker, R., Osterode, W. and Rudiger, H.W. (2002). Neurobehavioral effects of vanadium. *J. Toxicol. Environ. Health*, 65:677–683.

Berman, E. (1980). Toxic Metals and Their Analysis. Philadelphia: Hayden.

Bertrand, D. (1950) Survey of contemporary knowledge of biochemistry; II. The biochemistry of vanadium. *Bull. Am. Mus. Natl. Hist.* 94: 407–455.

Bracken, W.M., Sharma, R.P. and Elsner, Y.Y. (1985) Vanadium accumulation and subcellular distribution in relation to vanadate induced cytotoxicity in vitro. *Cell Biol. Toxicol.* 1: 259–268.

Byerrum, R.U., Eckardt, R.E., Hopkins, L.L., Libsch, J.F., Rostoker, W., Zenz, C., Gordon, W.A., Mountain, J.T., Hicks, S.P. and Boaz, T.D. (1974) Vanadium. Washington DC, National Academy of Sciences, pp. 1–117.

Byrne, A.R. and Kosta, L. (1978) Vanadium in foods and in human body fluids and tissues. *Sci. Total Environ.* 10:17–30.

Cam, M.C., Brownsey, R.W. and McNeill, J.H. (2000 o 1997) Mechanisms of vanadium action: insulin-mimetic and insulin-enhancing agensts? *Can. J. Physiol. Pharmacol.* 78: 829–847.

Campbell, C.A., Peet, M. and Ward, N.I. (1988) Vanadium and other trace elements in patients taking lithium. *Biol. Psychiat.* 24:775–781.

Cantley, L.C.Jr., Josephson, L., Warner, R., Yanagisawa, M., Lechene, C. and Guidotti, G. (1977) Vanadate is a potent (Na, K)-ATPase inhibitor found in ATP derived from muscle. *J. Biol. Chem.* 252:7421–7423.

Colin-Barenque, L., Avila-Costa, M.R., Sánchez, I., López, I., Niño-Cabrera, G., Pasos, F., Delgado, V. and Fortul, T.I. (2003) Muerte neuronal en el bulbo olfatorio de ratón inducida por exposición subaguda y crónica de vanadio. In: Proceedings of the XLVI

Congreso Nacional de la Sociedad de Ciencias Fisiológicas, A.C., Aguascalientes, Ags. México.

Conri, C., Simonff, M., Fleury, B. and Moreau, F. (1986). Variations in serum vanadium levels during treatment of mental depression. *Biol. Psychiatry*, 21:1335–1339.

Crans, D.C., Smee, J.J., Gaidamauskas, E. and Yang, L. (2004) The chemistry and biochemistry of vanadium and the biological activities exerted by vanadium compounds. *Chem. Rev.* 104: 849-902.

Fantus, I.G., Deragon, G., Lai, R. and Tang, S. (1995) Modulation of insulin action by vanadate: evidence of a role for phosphotyrosine phosphatase activity to alter cellular signaling. *Mol. Cell Biochem.* 153: 103–112.

Fawcett, J.P., Farquhar, S.J., Thou, T. and Shand, B.I. (1997) Oral vanadyl sulphate does not affect blood cells, viscosity or biochemistry in humans. *Pharmacol. Toxicol.* 80: 202–206.

Fiala, J.C., Spacek, J. and Harris, K.M. (2002) Dendritic Spine Pathology: Cause or Consequence of Neurological Disorders? *Brain Res. Rev.* 39: 29–54.

Grady, C.L., McIntosh, A.R., Horwitz, B., Maisog, J.M., Ungerleider, L.G., Mentis, M.J., Pietrini, P., Schapiro, M.B. and Haxby, J.V. (1995) Age-Related Reductions in Human Recognition Memory Due to Impaired Encoding. *Science*, 269: 218-221.

Guidotti, T.L., Audette, R.J. and Martin, C.J. (1997) Interpretation of the trace metal analysis profile for patients occupationally exposed to metals. *Occup. Med.* (*Lond.*). 47: 497–503.

Hamel, F.G. and Duckworth, W.C. (1995) The relationship between insulin and vanadium metabolism in insulin target tissues. *Mol. Cell Biochem.* 153: 95–102.

Hansen, T.V., Aaseth, J. and Alexander, J. (1982) The effect of chelating agents on vanadium distribution in the rat body and on uptake by human erythrocytes. *Arch. Toxicol.* 50: 195-202.

Harland, B.F. and Harden-Williams, B.A. (1994) Is vanadium of human nutritional importance yet? *J. Am. Diet Assoc.* 94:891–894.

Hathcock, J.N., Hill, C.H. and Tore, S.B. (1966) Uncoupling of oxidative phosphorylation by vanadate. *Can. J. Biochem.* 44: 983–988.

Hirano, S. and Suzuki, K.T. (1996) Exposure metabolism and toxicity for rare earths and related compounds. *Environ. Health Perspect.* 104: 85–95.

Hudson, T.G.F. (1964) Vanadium: Toxicology and biological significance. In: Browning, E., ed. Elsevier Monographs on Toxic Agents, Amsterdam, Elsevier Publishing Company, pp. 67–78.

Huyer, G., Liu, S., Kelly, J., Moffat, J., Payette, P., Kennedy, B., Tsaprailis, G., Gresser, M.J. and Ramachandran, C. (1997) Mechanism of inhibition of proteintyrosine phosphatases by vanadate and pervanadate. *J. Biol. Chem.* 272: 843– 851.

IARC Monographs on the Evaluation of Carcinogenic Risks to Humans. (2006) *Vanadium Pentoxide*. Volume 86: 227-292.

Janiszewska, G., Lachowicz, L., Jaskolski, D. and Gromadzinska, E. (1994) Vanadium inhibition of human parietal lobe ATPases. *Int. J. Biochem.* 26: 551–553.

Jankovic, J. (2008) Parkinson's disease: clinical features and diagnosis. *J. Neurol. Neurosurg Psychiatry*, 79:368-376.

Kandel, E.R. (2001) The molecular biology of memory storage: a dialogue between genes and synapses. *Science*, 294: 1030 –1038.

Kiviluoto, M., Pyy, L. and Pakarinen, A. (1981) Clinical laboratory results of vanadium exposed workers. *Arch. Environ. Health*, 36: 109–113.

Kustin, K., McLeod, G.C., Gilbert, T.R. and Briggs, L.B.R.T. (1983) Vanadium and other metal ions in the physiological ecology of marine organisms. *Struct. Bonding*, 53: 139-160.

Lewis, C.E. (1959) The biological effects of vanadium. II. The signs and symptoms of occupational vanadium exposure. *Arch. Ind. Health*, 19: 497–503.

Li, J., Elberg, G., Crans, D.C. and Shechter, Y. (1996) Evidences for the distinct vanadyl(+4)-dependent activating system for manifesting insulin-like effects. *Biochemistry*, 35: 8314–8328.

Luo, Y., Umegaki, H., Wang, X., Abe, R. and Roth, G.S. (1998) Dopamine induces apoptosis through an oxidation-involved SAPK/JNK activation pathway. *J. Biol. Chem*. 273:3756–3764.

Madany, I.M. and Raveendran, E. (1992) Polycyclic aromatic hydrocarbons, nickel and vanadium in air particulate matter in Bahrain during the burning of oil fields in Kuwait. *Sci. Total Environ*. 116: 281–289.

Meish, H.U. and Bielig, H.J. (1980) Chemistry and biochemistry of vanadium. *Basic Res. Cardiol*. 75:413–417.

Mesholam, R.I., Moberg, P.J., Mahr, R.N. and Doty, R.L. (1998) Olfaction in neurodegenerative disease : A meta-analysis of Olfactory functioning in Alzheimer's and Parkinson's diseases. *Arch. Neurol*. 55: 84-90.

Meza, P., Colin-Barenque, L., Mondragón, A., Avila-Costa, M.R., Sánchez I, López I, Niño-Cabrera G, Pasos F, Delgado V. and Fortoul, T. (2003) Efecto de la inhalación subaguda y crónica de vanadio sobre la ultraestructura del cerebelo de ratón. In: Proceedings of the XLVI Congreso Nacional de la Sociedad de Ciencias Fisiológicas, A.C., Aguascalientes, Ags. México.

Michibata, H. and Sakurai, H. (1990) Vanadium in Biological Systems; N.D. Chasteen, Ed.; Kluwer Academic Publishers: Boston.

Michibata, H., Uyama, T., Ueki T. and Kanamori, K. (2003a) Vanadocytes, cells hold the key to resolving the highly selective accumulation and reduction of vanadium in ascidians. *Microsc. Res. Technol*. 56: 421-434.

Michibata, H., Yamaguchi, N., Uyama, T. and Ueki, T. (2003b) Molecular biological approaches to the accumulation and reduction of vanadium by ascidians. *Coord. Chem. Rev*. 237:41-51.

Mitchell, W.G. (1953) Influence of pH on toxicity of vanadium in mice. *Proc. Soc. Exp. Biol. Med*. 84:404–405.

Mondragón, A., Colin-Barenque, L., Meza, P., Avila-Costa, M.R., Sánchez, I., López, I., Niño-Cabrera, G., Pasos, F., Delgado, V., Ordóñez, J., Gutiérrez, A., Aley, P. and Fortoul, T. (2003) Efectos de la inhibición crónica de vanadio sobre la citología del bulbo olfatorio de ratón. In: Proceedings of the XLVI Congreso Nacional de la Sociedad de Ciencias Fisiológicas, A.C., Aguascalientes, Ags. México.

Mukherjee, B., Patra, B., Mahapatra, S., Banerjee, P., Tiwari, A. and Chatterjee, M. (2004) Vanadium--an element of atypical biological significance. *Toxicol. Lett*. 150: 135-143.

Myron, D.R., Givand, S.H. and Nielsen, F.H. (1977) Vanadium content of selected foods as determined by flameless atomic absorption spectroscopy. *J. Agric. Food Chem*. 25: 297–299.

National Toxicology Program (2002) Toxicology and Carcinogenesis Studies of Vanadium Pentoxide (CAS No. 1314-62-1) in F3344/N rats and B6C3F1 mice (Inhalation Studies) (Technical Report series No. 507; NIH Publication No. 03-4441) Research Triangle Park, NC.

Naylor, G.J., Smith, A.H.W., Bryce-Smith, D. and Ward, N.I. (1984) Tissue vanadium levels in manic-depressive psychosis. *Psychol. Med.* 14: 767–772.

Nechay, R. (1984). Mechanisms of action of vanadate. *Annu. Rev. Pharmacol. Toxicol.* 24:510–524.

Nielsen, F.G. (1991) Nutritional requirements for boron, silicon, vanadium, nickel, and arsenic: Current knowledge and speculation. *FASEB J.* 5: 2661–2667.

Nielsen, F.H. and Uthus, E.O. (1990) The essentiality and metabolism of vanadium. In: Chasteen ND. Ed. Vanadium in Biological Systems. Kluwer Academic Publishers: Boston.

Nriagu, J.P. (1998) Vanadium in the Environment, Part 2: Health Effects, JohnWiley and sons, New York, Chichester,Weinheim, Brisbane, Singapore, Toronto.

Ramandham, S., Heyliger, C., Gresser, M.J., Tracey, A.S. and McNeil, J.H. (1991) The distribution and half life for retention of vanadium in organs of normal and diabetic rats orally fed vanadium(IV) and vanadium(V). *Biol. Trace Elem. Res.* 130: 119–124.

Ponsen, M.M., Stoffers, D., Booij, J., van Eck-Smit, B.L., Wolters, E.Ch. and Berendse H.W. (2004) Idiopathic hyposmia as a preclinical sign of Parkinson's disease. *Ann. Neurol.* 56: 173–81.

Poucheret, P., Verma, S., Grynpas, M.D. and McNeill, J.H. (1998) Vanadium and diabetes. *Mol. Cell Biochem.* 188: 73–80.

Rehder, D. (1999) The coordination chemistry of vanadium as related to its biological functions. *Coord. Chem. Rev.* 182: 297-322.

Rehder, D. (2003) Biological and medicinal aspects of vanadium. *Inorg. Chem. Commun.* 6: 604-617.

Roshchin, A.V. (1967) Toxicology of vanadium compounds used in modern industry. *Hyg. Sanit.* 32: 345–352.

Sasi, M.M., Haider, S.S., Fakhri, M-El. and Ghwarsha, K.M. (1994) Microchromatographic analysis of lipids, protein and occurrence of lipid peroxidation in various brain areas of vanadium exposed rats: A possible mechanism of vanadium neurotoxicity. *Neurotoxicity*, 15: 413-420.

Setyawati, I.A., Thompson, K.H., Yeun, V.G., Sun, Y., Battell, M., Lyster, D.M., Vo, C., Ruth, T.J., Zeisler, S., McNeill, J.H. and Orvig, C. (1998) Kinetic analysis and comparison of uptake, distribution, and excretion of 48 V-labeled compounds in rats. *J. Appl. Physiol.* 84: 569–575.

Sharma, R.P., Coulombe, R.A. and Srisuchart, B. (1986). Effects of dietary vanadium exposure on levels of regional brain neurotransmitters and their metabolites. *Biochem. Pharmacol.* 35: 461–465.

Simonian, N.A. and Coyle, J.T (1996) Oxidative stress in neurodegenerative disease. *Annu. Rev. Pharmacol. Toxicol.* 36: 83-106.

Tlan Wang, Y. and Salter, M.W. (1994) Regulation of NMDA receptors by tyrosine kinases and phosphatases. *Nature*, 369: 233-235.

Van Pee, K.H., Keller, S., Wage, T., Wynands, I., Schnerr, H. and Zehner, S. (2000) Enzymatic halogenation catalyzed via a catalytic triad and by oxidoreductases. *Biol. Chem.* 381: 1-5.

Wenning, R. and Kirsch, N. (1988). Vanadium. In: Seiler HG, Sigel H, Sigel A. eds. Handbook on Toxicity of Inorganic Compounds. New York: Marcel Dekker, pp. 749–765.

WHO (1988). Vanadium. In: Environmental Health Criteria. No. 81. Technical Report Series, Geneva: WHO.

Witkowska, D. and Brzezinski, J. (1979). Alteration of brain noradrenaline, dopamine and 5-hydroxytryptamine levels during vanadium poisoning. *Pol. J. Pharmacol. Pharm.* 31: 393–398.

Witkowska, D. and Brzezinski, J. (1983). Effect of metavanadate on the uptake and release of noradrenaline in rat brain cerebral cortex slices. *Toxicol. Lett.* 17: 223–231.

Wyers, H. (1946) Some toxic effects of vanadium pentoxide. *Br. J. Ind. Med.* 3: 177–182.

Zhao, Z., T. Zhongjia, C.D., Diltz, M., You, and Fischer, E.H. (1996) Activation of mitogen-activated protein (MAP) kinase pathway by pervanadate, a potent inhibitor of tyrosine phosphatases. *J. Biol. Chem.* 271: 22251–22255.

Zhen, X., Torres, C., Cai, G. and Friedman, E. (2002) Inhibition of protein tyrosine/mitogen-activated protein kinase phosphatase activity is associated with d2 dopamine receptor supersensitivity in a rat model of parkinson's disease. *Mol. Pharmacol.* 62: 1356–1363.

In: Vanadium
Editors: V. N. Baranova and A. V. Fortunatov
ISBN 978-1-62081-600-4

Chapter 7

SOFT SOLUTION PROCESSING OF NANOSCALED LITHIUM VANADIUM OXIDES AS CATHODE MATERIALS FOR RECHARGEABLE LITHIUM ION BATTERIES

Hao Wang*, HaiYan Xu and Hui Yan
The College of Materials Science and Engineering,
Beijing University of Technology, Beijing, China

ABSTRACT

Lithium vanadium oxides have been extensively studied because of their possible application as a cathode material for rechargeable lithium batteries. Due to their low cost, they are one of the promising substitutes for the expensive $LiCoO_2$ cathode presently commercially used. Lithium vanadium oxides including γ-LiV_2O_5 and LiV_3O_8 have been prepared by soft solution methods in this study.

In the first part of this work, γ-LiV_2O_5 nanorods have been prepared directly by a simple solvothermal method using ethanol as a solvent, which also serves as a reducing agent. The γ-LiV_2O_5 nanorods with diameters of 30-40 nm obtained at 160 °C shows a larger capacity of 259 mAh/g in the range of 1.5 - 4.2 V, and its capacity remained 199 mAh/g after 20 cycles.

In the second part, LiV_3O_8 nanorods have been obtained by a novel hydrothermal-based two-step method. The LiV_3O_8 sample treated at 300 °C shows a poor crystallinity while a specific capacity of 302 mAh/g in the range of 1.8 - 4.0 V, and its capacity remained 278 mAh/g after 30 cycles.

It indicates that the lithium vanadium oxide nanorods prepared by the above methods have potentiality to be used as cathode material in rechargeable lithium ion batteries.

* Corresponding author: E-mail: haowang@bjut.edu.cn.

INTRODUCTION

Worldwide research and development is in progress for rechargeable lithium ion batteries because of its wide application in portable electronic goods, electric vehicle systems and dispersed-type energy storage system. Much of this effort is focused on developing the cathode material for lithium ion batteries because the performance and cost of the batteries are often decided by the properties of the cathode material. In the last 20 years, vanadium oxides have been widely investigated as cathode materials in lithium ion batteries due to their low cost, low toxicity, high specific energy, and long cycle life. Lithium ion can react with V_2O_5 to form various compounds such as LiV_3O_8 [1-5], LiV_2O_5 (LiV_2O_5 has many modifications including α-, β-, ω-, γ-LiV_2O_5) [6-10], LiV_2O_4 [11], $Li_{0.6}V_{2-\delta}O_{4-\delta}$ [12] etc. Among them, γ-LiV_2O_5 and LiV_3O_8 were most intensively studied [1-10, 13-15].

The preparation method has significant influences on the electrochemical behavior of γ-LiV_2O_5 and LiV_3O_8, therefore a great deal of work has been done on their preparation to improve their electrochemical performance. In this study, soft solution methods including a solvothermal route and a modified hydrothermal-based method have been performed to prepare γ-LiV_2O_5 and LiV_3O_8 nanorods, respectively.

$LI_{1+X}V_3O_8$

Introduction

LiV_3O_8, as a species of the lithiated vanadate family, has very attractive characteristics such as high specific energy, good rate capacity, and long cycle life. The structure of LiV_3O_8 can be shown as in Figure 1. [16] It has a monoclinic structure and the space group in $P2_1/m$. The structure consists of VO_6 octahedra and VO_5 distorted trigonal bipyramids and is built up by sharing the edges and corners of the trigonal bipyramids and the octahedra to form a zigzag ribbon. Lithium can be localized in both of two kinds of vacant sites named octahedral and tetrahedral sites (Figure 1), in which the original combined lithium atoms occupy octahedral coordinated sites and the intercalated ones occupy tetrahedral sites, respectively.

It is well known that the preparation methods and post-treatments have significant influences on the electrochemical properties of LiV_3O_8. The conventional method was high temperature melting in which LiV_3O_8 was produced by reaction between Li_2CO_3 and V_2O_5 at 680 ^{o}C [17, 18]. This method has met difficulty to control the composition and homogeneity of the final products. Meanwhile, the product LiV_3O_8 had a low capacity of 180 mAh/g in the range of 1.8-4.0 V. Afterwards, many improved solution methods were proposed [2, 4, 19-21]. The solution method does not need a high reaction temperature, and the products could reach a high homogeneity and high capacity.

As one of the solution methods, hydrothermal method has been extensively used for the synthesis of inorganic compounds. Chirayil et al. had obtained a new layered lithium vanadium oxide $Li_xV_{2-\delta}O_{4-\delta}\cdot H_2O$ via hydrothermal method using tetra-methyl ammonium as template [11]. Oka et al. had prepared AV_3O_8 (A=K, Rb, and Cs) by the hydrothermal treatment of V_2O_5 and ANO_3 solutions at 250 ^{o}C. The results showed that RbV_3O_8 had the known structure of the KV_3O_8 and CsV_3O_8 compounds [22, 23]. However, attempts to make LiV_3O_8 and NaV_3O_8 hydrothermally were unsuccessful.

Some reports have also noticed the importance of crystallinity on the electrochemical properties of LiV_3O_8. It was shown that the poorly crystallized LiV_3O_8 or LiV_3O_8 modified by ultrasonic treatment or hydrothermal introduction of small amounts of inorganic compounds such as H_2O, CO_2 and NH_3 achieved an improved electrochemical performance [3, 5]. It is known that the electrochemical behavior of cathode materials strongly depends on the particle size: the bigger the particle size, the higher the cell polarization and the lower the cell capacity [8]. In the present paper, LiV_3O_8 nanorods have been prepared by a novel method, which is based on the hydrothermal reaction, the experimental procedure and results are described below.

Experimental Section

Synthesis and Characterization of LiV_3O_8

The starting materials were analytically pure LiOH, V_2O_5 and $NH_3{\cdot}H_2O$ (1 mol/L). First, sotoichiometric LiOH and V_2O_5 (Li : V = 1 : 3, molar ratio) were blended in the deionized water. LiOH was dissolved completely and part of V_2O_5 was dissolved. When $NH_3{\cdot}H_2O$ was added to the above mixture, V_2O_5 was dissolved completely into solution. The pH of the solution was 9.

The resultant dark green solution was then transferred to a 50 ml Teflon lined autoclave. The autoclave was sealed and heated at 160 °C under auto-generated pressure for 12 h. After hydrothermal treatment, a colorless clear solution whose pH changed to 7 was obtained. This solution was dried in air at 100 °C to evaporate the water till an orange gel was prepared. The gel was then heat-treated at different temperatures in the range of 300-600 °C for 12 h.

The gel was characterized by thermo-gravimetric analysis (TGA) using a Model STA 449C (Germany, NETZSCH-Gerätebau GmbH Thermal Analysis). The structure of the heat-treated products was examined by X-ray diffractometry (XRD, Japan Rigaku D/Max-3C) using CuKα radiation (λ = 1.5405Å). The morphology was investigated by transmission electron microscopy (TEM, Model Hitachi H-700H, 200 KV). Fourier transform infrared (FTIR) absorption spectra were obtained by using a Nicolet Magna-IR 560 spectroscopy.

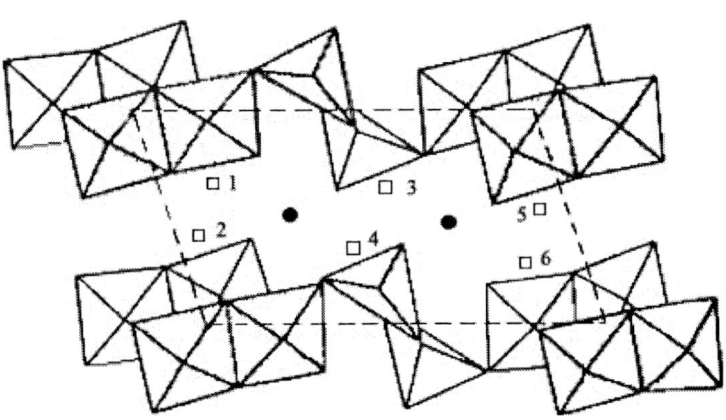

Figure 1. Structure model of LiV_3O_8. Key: (•) octahedral and (□) tetrahedral.

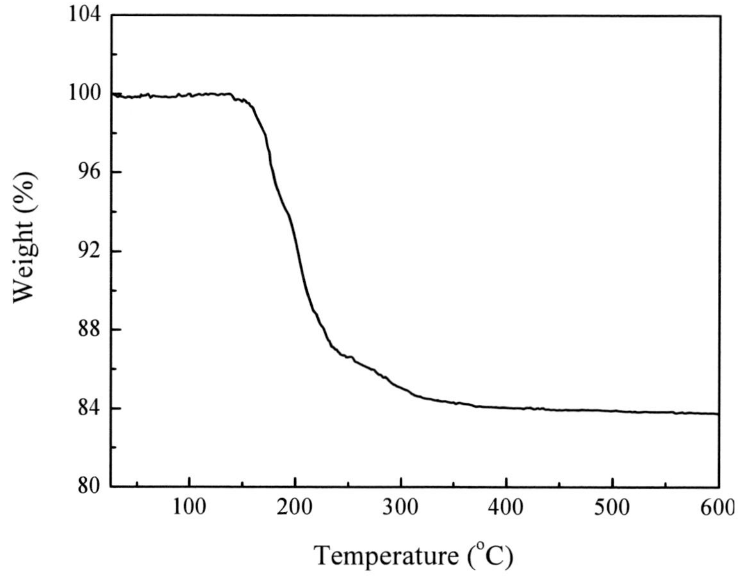

Figure 2. TGA curve of the precursor gel.

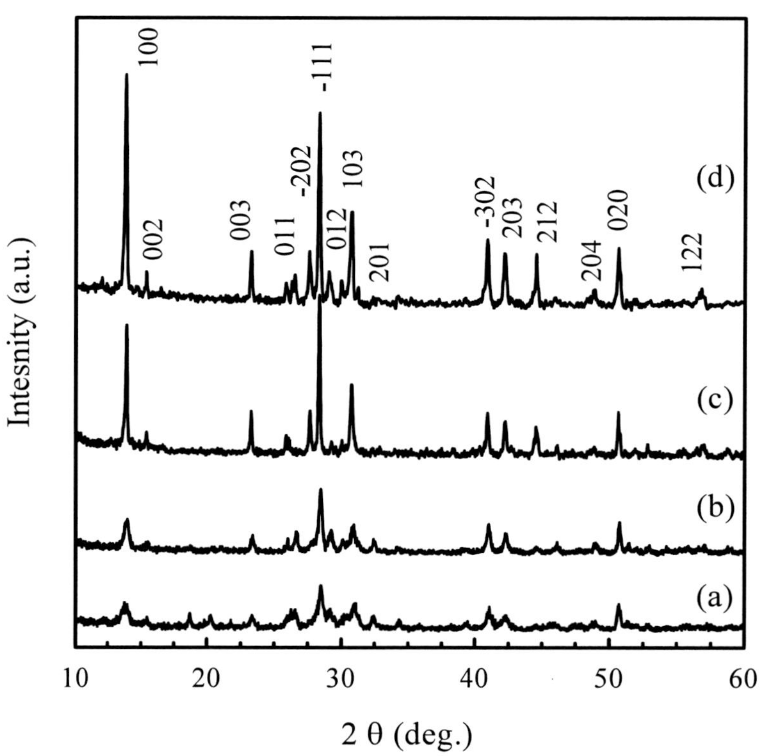

Figure 3. The XRD patterns of the samples heat treated at (a) 300, (b) 350, (c) 400, (d) 600 °C.

Electrochemical Measurements

Electrochemical characterization of the products was performed in cells with metallic lithium as the negative electrode and a liquid organic electrolyte [$LiPF_6$ in a volume ratio of 50:50 mixture of ethylene carbonate (EC) and diethyl carbonate (DEC), absorbed in porous polypropylene separators, Celgard 2400]. The cathode was a mixture of the active material, acetylene black and poly (tetrafluoroethylene) (PTFE) in a weight ratio of 80:10:10. The cells were assembled in an argon-filled dry box. Charge-discharge tests were carried out at a constant current density of 0.3 mA/cm^2 in a range of 1.8-4.0 V. All the tests were performed at room temperature.

Results and Discussion

The TGA result of the precursor gel which was derived by hydrothermal reaction and subsequent evaporation of water is illustrated in Figure 2. It can be seen that in the range of 140 ~ 320 °C, the weight loss is about 15 %. Above 320 °C, the weight remains stable up to 600 °C. The weight loss process can be divided into two stages. The first weight loss begins at 140 °C and ends at 230 °C with a weight loss of 13%, which mainly resulted from the evaporation of NH_3 and water. The second weight loss occurred in the range of 250 ~ 320 °C with a weight loss of 2 %, which was caused by the de-intercalation of some strongly-bound water.

TGA Result

The XRD and the Structure of LiV_3O_8

Figure 3 shows the XRD patterns of the products heated at 300, 350, 400 and 600 °C. These XRD patterns reflect the structural variation during the treatment process. With an increase in heat temperature the intensity of peaks becomes stronger and the full width of half maximum intensity (FWHM) decreases, which indicates that the crystallinity becomes higher.

In addition, it is noted that there is an obvious difference among the four XRD patterns. The relative intensity of (100) peaks at around 13.86° becomes stronger with increasing heat temperature. In the 600 °C diffraction pattern, the intensity of (100) peak is stronger than any other peaks. While in the 300 °C diffraction pattern, the relative intensity of (100) peak has much decreased. It suggests that the particle shape of LiV_3O_8 depends on the heat temperature, and higher heat temperature favors the preferential ordering of crystallites. It is known that the intercalation process of Li^+ ion between the layers of the cathode is a diffusion process. Therefore the preferential ordering which would lead to a long path for Li^+ ion is not advantageous to intercalation [2].

The Morphology of the As-Synthesized LiV_3O_8

The TEM micrographs of the synthesized LiV_3O_8, at different heat temperatures, are demonstrated in Figure 4. It can be seen that the heat treatment has caused a change in the LiV_3O_8 crystallinity and morphology. The sample treated at 300 °C (Figure 4a) consists of an agglomeration of small rods. The shapes of the rods are not well recognized, indicating they are relatively poorly crystallized.

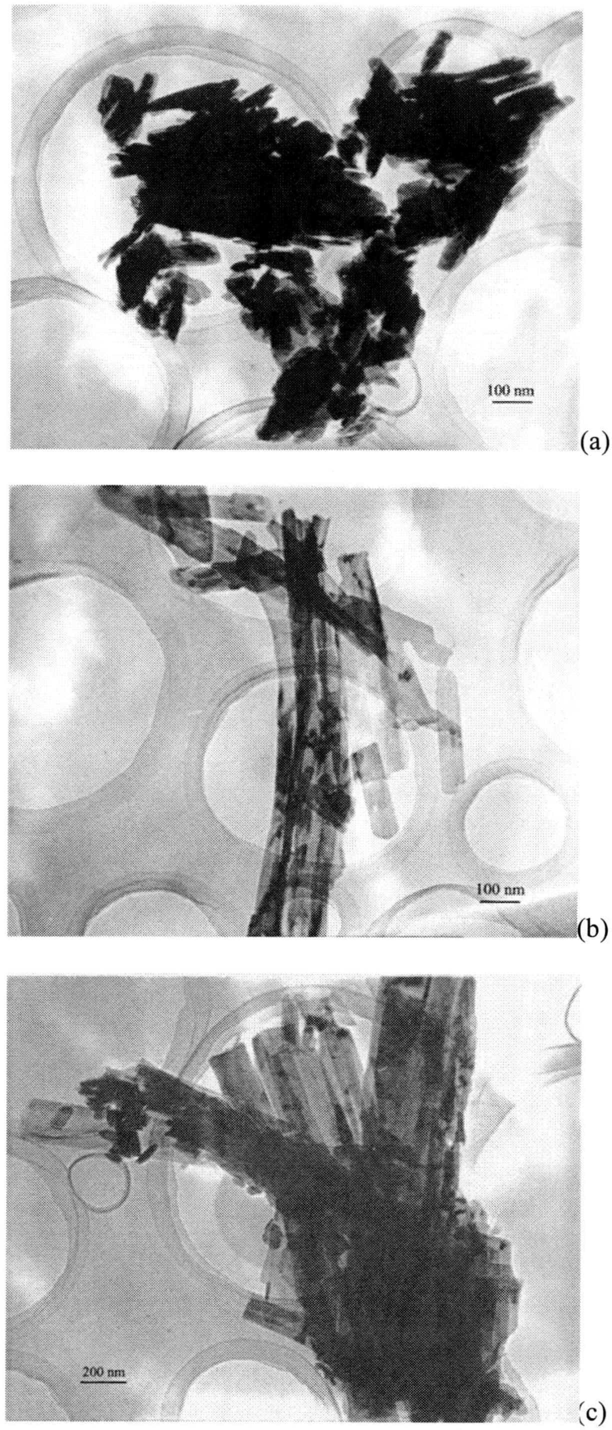

Figure 4. TEM micrographs of the samples heat treated at (a) 300, (b) 350, (c) 400 °C.

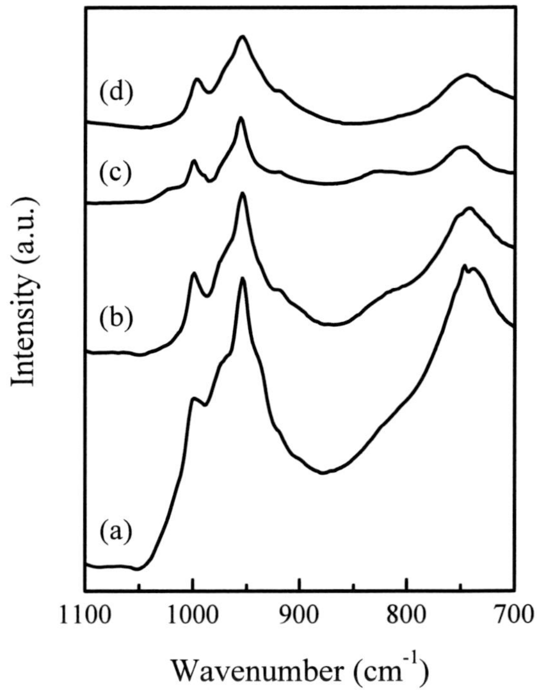

Figure 5. The FTIR spectra of the samples heat treated at (a) 300, (b) 350, (c) 400, (d) 600 °C.

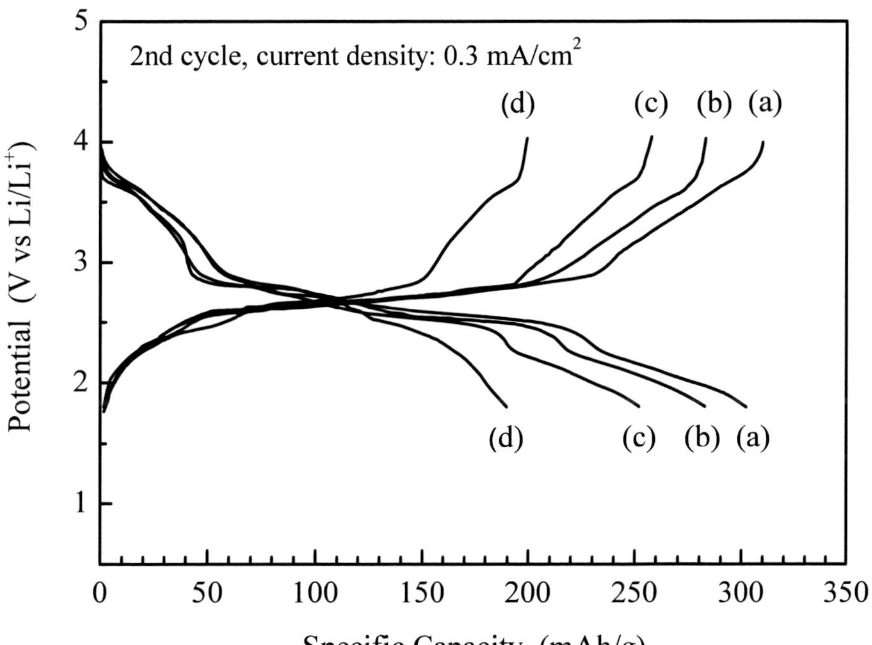

Figure 6. Charge-discharge curves of the samples heat treated at (a) 300, (b) 350, (c) 400, (d) 600 °C. Current density: 0.3 mA/cm^2.

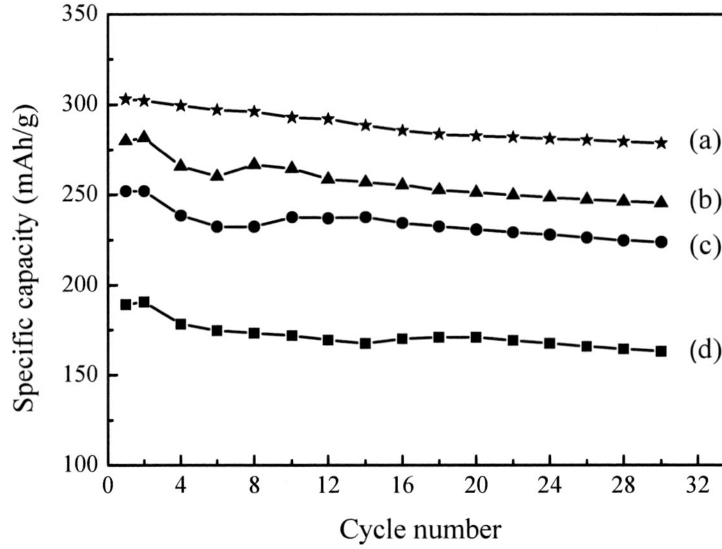

Figure 7. The cycle performance of the cells with LiV_3O_8 heat treated at (a) 300, (b) 350, (c) 400, (d) 600 °C as cathode active material. Current density: 0.3 mA/cm^2. Voltage window: 1.8-4.0 V.

The diameters of the rods are about 40 nm and the lengths are mostly less than 600 nm. Figure 4b is the micrograph of the sample treated at 350 °C. Comparing to the 300 °C sample, the 350 °C sample is well rod-shaped. The diameters of the rods are about 70 nm while the lengths are diverse in the 0.5-2 μm range. The 400 °C sample which is shown in Figure 4c consists of wider (>150 nm) and longer rods. The sizes of the particles treated at 600 °C are much larger than that of the 400 °C sample, the micrograph of which is not shown here. These results are in agreement with the XRD data. With the increase in temperature, the particles of the products become larger and more crystallized.

FTIR of the As-Synthesized LiV_3O_8

FTIR spectra of LiV_3O_8 obtained at different heat temperatures are shown in Figure 5. For the four curves, it can be seen that the FTIR absorption in the spectral region of 700 ~ 1100 cm^{-1} is dominated by bands at 995, 954 and 744 cm^{-1}, respectively. This data is compared with LiV_3O_8 synthesized from solid state reactions, where major FTIR bands corresponding to V = O and V-O-V vibrations are located at 996, 954 and 746 cm^{-1}, respectively [24]. The sample treated at 300 °C gives the same FTIR responses as that of crystalline LiV_3O_8, suggesting that it is almost iso-structural with the latter.

Electrochemical Properties of the As-Synthesized LiV_3O_8

The charge-discharge curves at the second cycle of cells from products heat treated at 300, 350, 400, 600 °C, respectively, are illustrated in Figure 6. It is found that the specific capacities of the samples decrease with increasing the heat temperature. The discharge capacity for the 300 °C sample is 302 mAh/g in the range of 1.8 - 4.0 V, which is much higher than that of 600 °C sample (190 mAh/g). Notably, the value of 302 mAh/g is considerably higher than the capacities of 220 ~ 274 mAh/g of LiV_3O_8 synthesized by other solution methods [2, 4, 19-21], and exceeds the capacity of about 280 mAh/g of hydrothermally and ultrasonically treated LiV_3O_8 [3, 5]. From the XRD data we have known that the sample treated at higher temperature has higher crystallinity. The above results also suggest that the well-crystallized sample does have poor specific capacity, which is in agreement with the previous studies [2-5, 19-21].

In Figure 7, the discharge capacities are shown as a function of cycle number. It can be seen that the samples synthesized via this novel method show better capacity retention. The capacity of the 300 °C samples is 278 mAh/g after 30 cycles. Meanwhile, the 300 °C sample shows better cycling behavior compared with the samples treated at higher temperature.

Conclusion

In this study, a novel method, which is based on the hydrothermal reaction has been performed to synthesize LiV_3O_8 nanorods. This hydrothermal reaction did not directly yield a solid product. A gel, which was used as the precursor for the post heat treatment was obtained after evaporation of the hydrothermal treated solution. The gel was homogeneous and ultrafine, which should be the reason why the LiV_3O_8 nanorods could be obtained in this study. Heat treatment at different temperatures influenced the particle size and crystallinity of the products, which consequently affected their electrochemical performance. The sample treated at 600 °C shows a good crystallinity and a low discharge capacity. In contrast, the sample treated at 300 °C shows a poorer crystallinity while a better capacity of 302 mAh/g in the range of 1.8 - 4.0 V, and its capacity remained 278 mAh/g after 30 cycles.

γ-LiV_2O_5

Introduction

γ-LiV_2O_5, as one of the promising cathode materials for rechargeable lithium batteries, has attracted tremendous research interest in recent years [6, 9, 10]. Orthorhombic γ-LiV_2O_5 has a layer structure with lithium ions between the layers. In this structure, there are two crystallographic vanadium sites that form two kinds of zigzag chains, the shaded and the white zigzag chains, as shown in Figure 8 [25].

The valence states of the vanadium ions were inferred from the results of structural analysis to be V^{4+} for the shaded zigzag chains and V^{5+} for the white zigzag chains [25]. Within the layers, $V^{4+}O_5$ (shaded) zigzag chains are linked to $V^{5+}O_5$ (white) zigzag chains by corner sharing, as shown in Figure 8.

In particular, a lot of work has been devoted to the synthesis of γ-LiV_2O_5, because it is very difficult to synthesize this compound, in which there are two valence states (V^{5+} and V^{4+}) of vanadium co-existing. So far, solid-state reaction routes[2] and a procedure developed by Murphy et al. [7] were the two commonly used preparation method for the synthesis of γ-LiV_2O_5. By solid-state reaction routes, the mixture of $LiVO_3$, V_2O_5, and VO_2 or V_2O_3 was heated at above 600 °C under vacuum or nitrogen protected atmosphere in order to control the valence balance of V^{4+} and V^{5+}. This calcinating method is usually faced with difficulty to control the homogeneity and particle size of the final products.

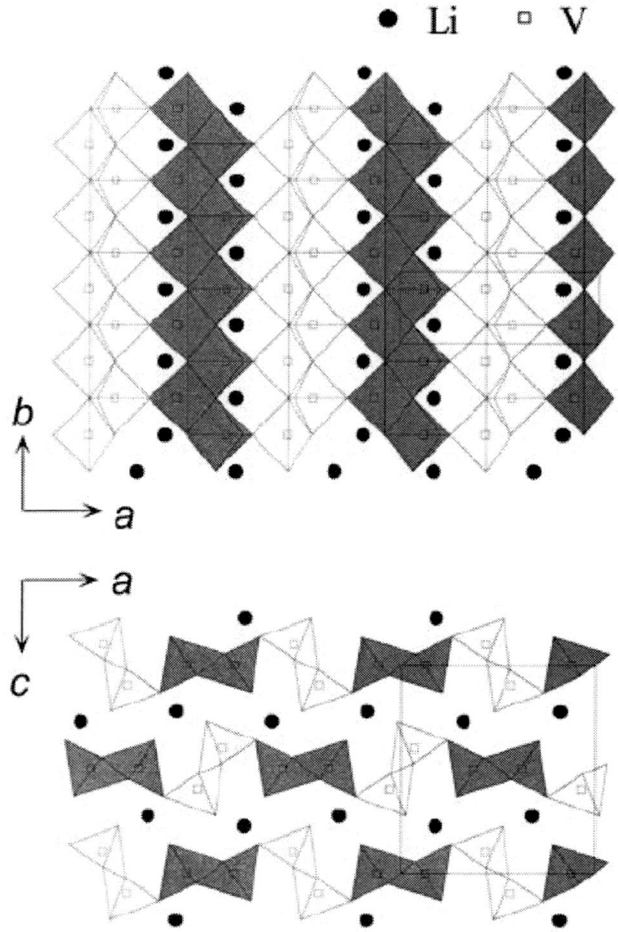

Figure 8. Schematic crystal structure of γ-LiV_2O_5 projected onto the orthorhombic *a-b* plane and *a-c* plane. The filled circles represent Li^+ ions. The squares represent V atoms. The white and shaded square pyramids show two kinds of VO_5 pyramids (two crystallographic vanadium sites).

It has been shown that the electrochemical behavior of cathode materials strongly depends on the particle size: the bigger the particle size, the higher the cell polarization and the lower the cell capacity [8]. The synthetic method developed by Murphy et al. includes two steps: firstly, δ-LiV_2O_5 was synthesized through the reduction of V_2O_5 with appropriate amount of lithium iodide in acetonitrile (CH_3CN) at ambient temperature; secondly, the as-prepared δ-$Li_xV_2O_5$ was heated at 350 °C under argon to obtain γ-LiV_2O_5. Such route has demonstrated the feasibility of preparing γ-LiV_2O_5 as a cathode material for rechargeable lithium ion batteries. However, the high cost and sensitivity to moisture of LiI and CH_3CN make it not viable as a low cost and low toxicity preparation method for cathode materials.

We have proposed a simple and mild solvothermal method for the synthesis of γ-LiV_2O_5. In this process, γ-LiV_2O_5 nanorods were synthesized directly from the solvothermal reaction of V_2O_5, LiOH and ethanol at 160-200 °C in an autoclave. Ethanol was employed as a solvent as well as a reducing agent. Compared with the two commonly used preparation methods for γ-LiV_2O_5, the newly developed solvothermal method is cheaper and milder, for example, vaccum or argon/nitrogen protected atmosphere or post annealing is not necessary for this simple one-step process. Therefore, the solvothermal process seems to offer a potentially low-temperature, low-cost and environmentally friendly way of producing single-phase, uniform-particle size and fine-grained γ-LiV_2O_5 for rechargeable lithium batteries. In this paper, the effect of experimental parameters including temperature and time on the properties of γ-LiV_2O_5 was also investigated. The as-obtained products were further characterized by X-ray diffraction (XRD), Fourier transform infrared (FTIR) spectroscopy, X-ray photoelectron spectrscopy (XPS), and transmission electron microscopy (TEM). The electrochemical performance of γ-LiV_2O_5 nanorods with different size was studied.

Experimental Section

Synthesis and Characterization of γ-LiV_2O_5

In a 50 ml Teflon vessel, 0.02 mol of analytically pure LiOH and V_2O_5 were mixed in 40 ml of ethanol. Be well stirred, the Teflon vessel containing the mixture was put into a stainless steel autoclave. After the autoclave was kept at 160, 180, 200 °C under autogeneous pressure for 12-48 h, it was allowed to cool to room temperature naturally. The as-formed solid precipitate was filtered, washed with ethanol, and dried at 100 °C for 2 h.

For all products, XRD (Japan Rigaku D/Max-3C, Cu Kα radiation) was utilized to identify the produced phase. Fourier transform infrared (FTIR) absorption spectra were obtained by using a Nicolet Magna-IR 560 spectroscopy.

XPS measurements were performed on a PHI-5300/ESCA system with Mg Kα radiation as the exciting source, where the binding energies were calibrated by referencing the C 1s peak to reduce the sample charge effect. The morphology and particle sizes of the as-obtained γ-LiV_2O_5 were determined by TEM (Model Hitachi H-700 H, 200 KV).

Electrochemical Measurements

Electrochemical tests were performed in cells with metallic lithium as the negative electrode. The electrolyte was $LiPF_6$ in the mixture of ethylene carbonate (EC) and diethyl carbonate (DEC) (50:50), Celgard 2400 as separators. The positive electrode composites were made by mixing the active material, acetylene black and poly (tetrafluoroethylene) (PTFE) (80:10:10). All cells used in this study were assembled in an argon-filled dry box. Charge-discharge tests were carried out at a constant current density of 0.3 mA/cm^2 in a range of 1.5 - 4.2 V. All tests were performed at room temperature.

Results and Discussion

The XRD Results and the Kinetic Processing of the Formation of γ-LiV_2O_5

Figure 9 shows the XRD patterns of the black products solvothermally synthesized at 160, 180, 200 °C for 24 h. As can be seen, all the obtained products have almost the same XRD patterns; which can be assigned to an orthorhombic structure γ-LiV_2O_5 according to JCPDS cards No. 018-756. It is suggested the formation of γ-LiV_2O_5 without any trace of impurities such as V_2O_5, Li_3VO_4 and $LiVO_2$ by this solvothermal route.

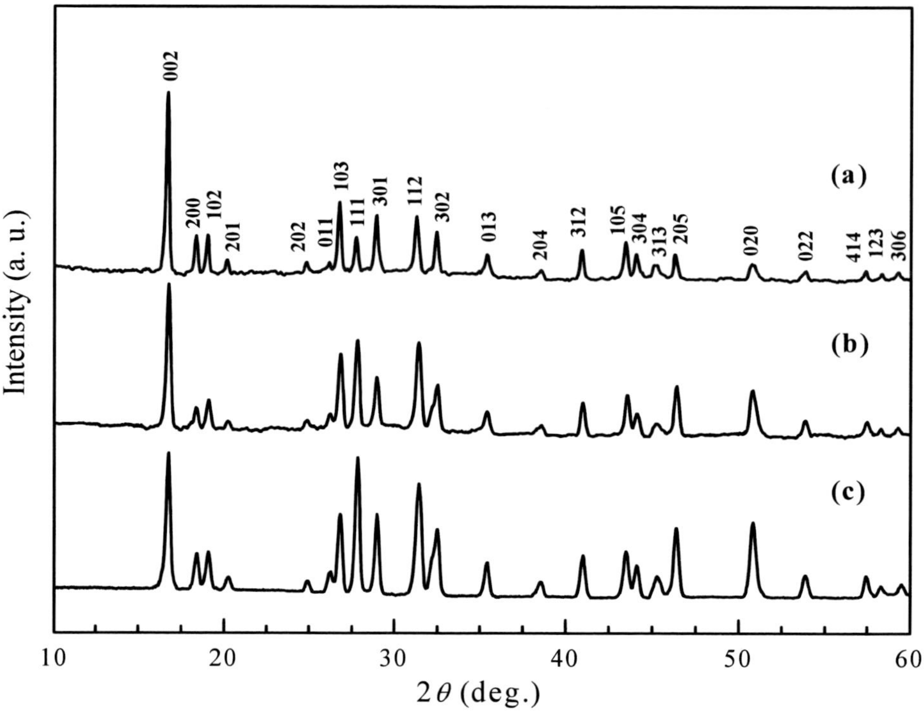

Figure 9. The XRD patterns of the samples synthesized at (a) 160, (b) 180, (c) 200 oC for 24 h.

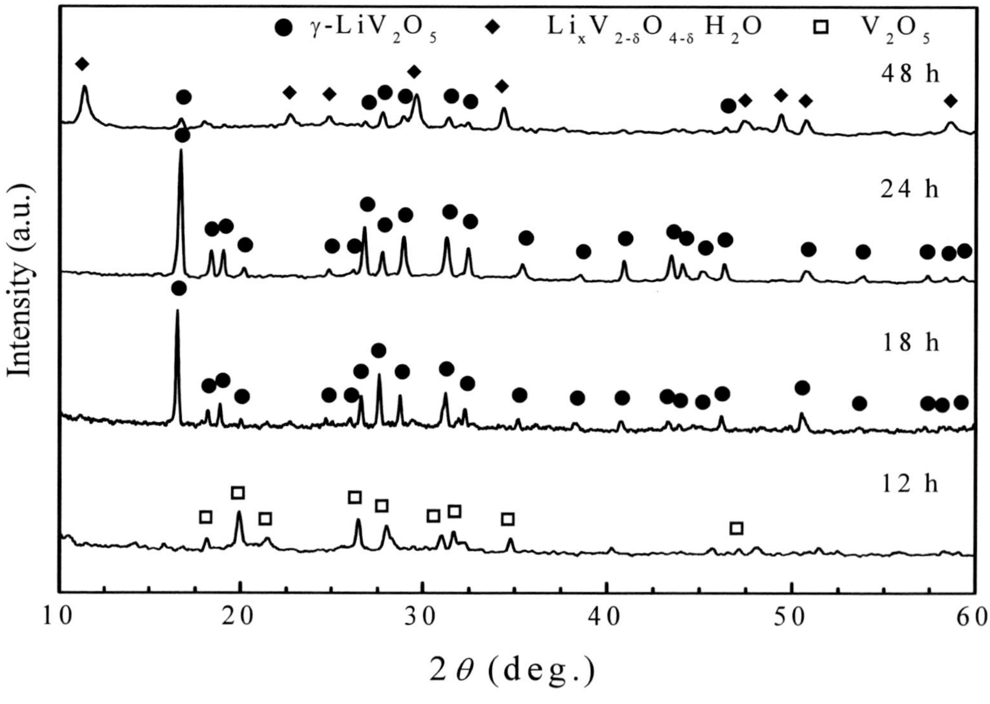

Figure 10. The XRD patterns of the samples synthesized at 160 °C for (a) 12, (b) 18, (c) 24, (d) 48 h.

Figure 10 shows the XRD patterns of the products solvothermally synthesized at 160 °C for 12-48 h. It is obvious that the variation of reaction time causes a change in the XRD patterns of the resulting products. For 12 h, the obtained product mainly consists of V_2O_5, in which the average valence of vanadium is +5; For 18-24 h, the obtained product is pure γ-LiV_2O_5, in which the average valence of vanadium is + 4.5. When the reaction time is further prolonged to 48 h, the obtained product is chiefly $Li_xV_{2-\delta}O_{4-\delta}\cdot H_2O$ [11] with a minor amount of γ-LiV_2O_5, in which the average valence of vanadium is lower than +4.

The above results indicate that when the reaction temperature is fixed at 160 °C, the reaction time plays an important role in determining the phases of products. We have also investigated the effect of the reaction time on the phase(s) of the resulting products at the temperatures of 180 and 200 °C, and obtained the similar results. It is suggested that, in this study, the formation of γ-LiV_2O_5 can be attributed to a process that is mainly controlled by kinetics. In our experiment, the synthesis of γ-LiV_2O_5 is carried out using a great excess of ethanol as a solvent through a solvothermal process. To the ethanol, a portion of it is oxidized (by V_2O_5) to aldehyde, which is experimentally confirmed through chemical analysis of the final solution mixture. From the above results, it can be seen that the ethanol acts not only as a solvent but also as a reducing agent in the synthesis of γ-LiV_2O_5 powders. The possible formation mechanism of γ-LiV_2O_5 through the solvothermal reaction of LiOH and V_2O_5 in ethanol medium is a reductive recombination pathway. Thus, the formation mechanism of γ-LiV_2O_5 through the reaction of LiOH, V_2O_5 and excess ethanol under the solvothermal condition can be expressed as the following:

$$2\ LiOH + 2\ V_2O_5 + CH_3CH_2OH \rightarrow 2\ \gamma\text{-}LiV_2O_5 + CH_3CHO + 2\ H_2O$$

If the reaction time is further prolonged, the obtained γ-LiV_2O_5 may be reduced to $Li_xV_{2-\delta}O_{4-\delta}\cdot H_2O$, and the reaction equation can be expressed as the following:

$$\gamma\text{-}LiV_2O_5 + CH_3CH_2OH \rightarrow Li_xV_{2-\delta}O_{4-\delta}\cdot H_2O + CH_3CHO$$

FTIR of the As-Synthesized γ-LiV_2O_5

Figure 11 shows the FTIR spectra of the as-synthesized γ-LiV_2O_5 at 160 °C for 24 h and the starting material V_2O_5. It can be seen, the FTIR absorption bands of γ-LiV_2O_5 are different from those of V_2O_5 in the spectral region of 400 ~ 1200 cm^{-1}. The FTIR spectrum of the as-synthesized γ-LiV_2O_5 (Figure 11 (a)) displays four major absorption bands at 1006, 952, 592, and 565 cm^{-1}, which are in agreement with the literature values. The absorption bands at 1006 and 952 cm^{-1} are attributed to V=O stretching vibrations, and the absorption bands at 592 and 565 cm^{-1} are attributed to V-O-V bending vibrations [26]. The FTIR spectrum of the starting material V_2O_5 (Figure 11 (b)) exhibits four major absorption bonds at 1025, 824, 600 and 473 cm^{-1} which have been associated with the V=O and V-O-V vibrations [26].

XPS of the As-Synthesized γ-LiV_2O_5

Figure 12 shows the XPS spectra of the starting material V_2O_5 and the as-synthesized γ-LiV_2O_5. It is obvious that the binding energy of V 2p for the as-synthesized γ-LiV_2O_5 is lower than that for the starting material V_2O_5. The binding energy of V $2p_{3/2}$ for the starting material V_2O_5 is 517.5eV, which agrees with the literature value of V_2O_5 [27]. The binding energy of V $2p_{3/2}$ for the as-synthesized γ-LiV_2O_5 is 516.7 eV, which is between those of V $2p_{3/2}$ for V_2O_5 and VO_2 (516.3 eV [27]). The average valence of vanadium in V_2O_5, γ-LiV_2O_5 and VO_2 is +5, +4.5 and +4, respectively.

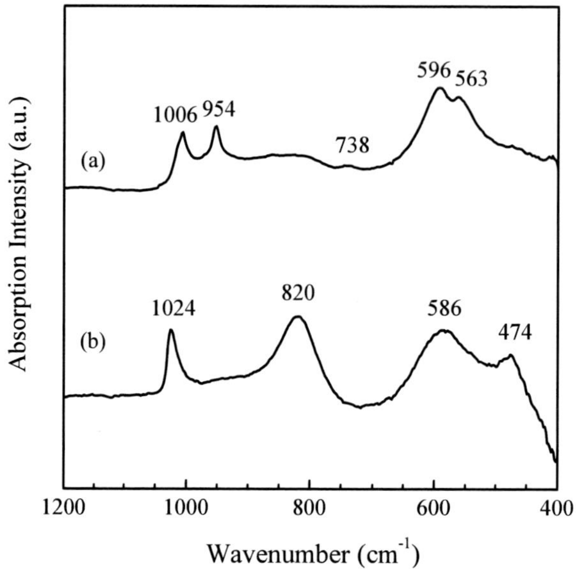

Figure 11. The FTIR spectra of (a) γ-LiV_2O_5 synthesized at 160 °C for 24 h, and (b) the starting material V_2O_5.

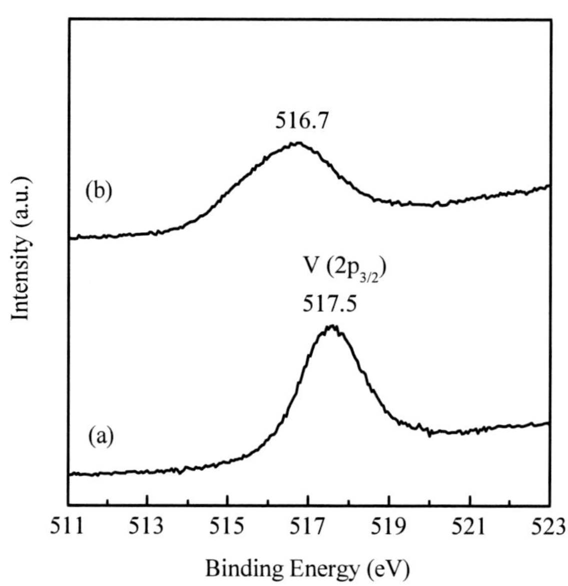

Figure 12. The XPS spectra of (a) γ-LiV_2O_5 synthesized at 160 °C for 24 h, and (b) the starting material V_2O_5.

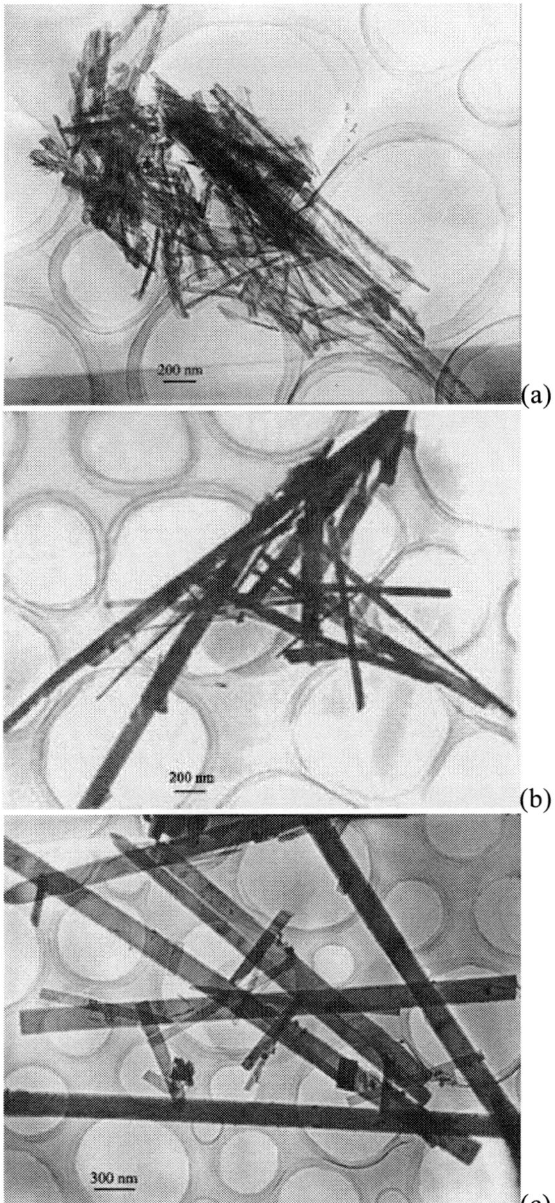

Figure 13. TEM micrographs of the γ-LiV_2O_5 synthesized at (a) 160, (b) 180, (c) 200 °C for 24 h.

The Morphologies of the As-Synthesized γ-LiV_2O_5

Figure 13 shows the TEM micrographs of the as-synthesized γ-LiV_2O_5 at 160,180, 200°C for 24 h. As can be seen, the solvothermal reaction temperature has an effect on the particles size of the as-synthesized γ-LiV_2O_5. With the increase in solvothermal reaction temperature, the as-synthesized γ-LiV_2O_5 particles become larger and even their size distribution becomes wider. The as-synthesized γ-LiV_2O_5 at 160 °C (Figure 13a) comprises nanorods with diameters of 30-40 nm and lengths of 0.4-2 μm; The as-synthesized γ-LiV_2O_5 at 180 °C (Figure 13b) consists of wider and longer rods with diameters of 50-120 nm and lengths of 0.7-3.5 μm; The as-synthesized γ-LiV_2O_5 at 200 °C for 24 h (Figure 13c) is composed of rods with diameters of 70-160 nm and lengths of 0.7-4 μm.

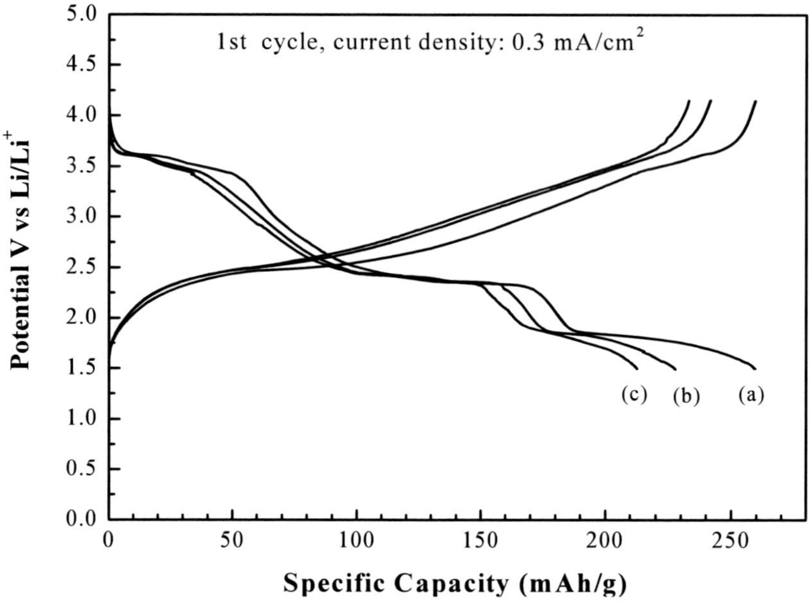

Figure 14. Charge-discharge curves of the cells with γ-LiV_2O_5 synthesized at (a) 160, (b) 180, (c) 200 °C for 24 h as cathode active material. Current density: 0.3 mA/cm^2.

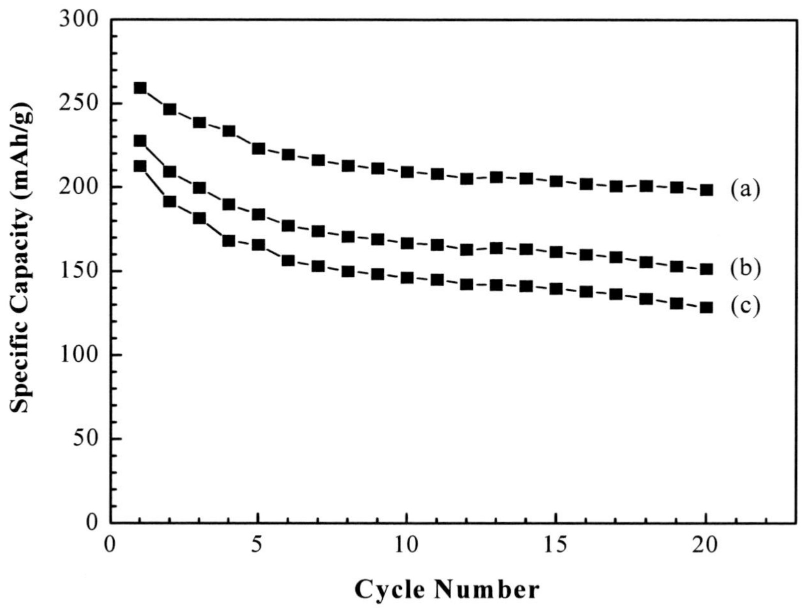

Figure 15. The cycle performance of the cells with γ-LiV_2O_5 synthesized at (a) 160, (b) 180, (c) 200 °C for 24 h as cathode active material. Current density: 0.3 mA/cm^2. Voltage window: 1.5-4.2 V.

Electrochemical Properties of the As-Synthesized γ-LiV$_2$O$_5$

Figure 14 shows the charge-discharge curves of the cell with the solvothermally-synthesized γ-LiV_2O_5 at 160, 180 and 200 °C for 24 h during the first cycle. It can be seen that the discharge capacity for the product synthesized at 160°C is 259 mAh/g in the range of 1.5 - 4.2 V, which is much higher than that of the product synthesized at 200 °C (218 mAh/g) and 180 °C (228 mAh/g). From the TEM results, we know that the solvothermal-synthesized γ-LiV_2O_5 at 200 °C has a larger size than that of the product obtained at 160 °C. Thus, it is reasonable to infer that the product with smaller particle size has a higher specific capacity, which is also consistent with the previous report [5].

Figure 15 shows the cycle performance of the cell with the as-synthesized γ-LiV_2O_5 at 160, 180 and 200 °C for 24 h. It can be seen that the solvothermally synthesized γ-LiV_2O_5 shows better capacity retention. In the first 6 cycles, the specific capacities of three cells reduce quickly, and then maintain stable in the subsequent cycles. The specific capacity for the cell with γ-LiV_2O_5 synthesized at 160 °C is 199 mAh/g after 20 cycles. The specific capacity for the cell with γ-LiV_2O_5 synthesized at 180 °C and 200 °C is 152 mAh/g and 129 mAh/g after 20 cycles, respectively. The above results indicate that the γ-LiV_2O_5 nanorods synthesized by this solvothermal method have potentiality to be used as a cathode material in rechargeable lithium cells.

CONCLUSION

In summary, LiV_3O_8 nanorods have been obtained by a novel two-step method which was based on the hydrothermal reaction. The as-prepared LiV_3O_8 nanorods showed a good charge-discharge and cycle performance. Heat treatment at different temperatures influenced the particle size and crystallinity of the products, which consequently affected their electrochemical performance. The sample treated at 300 °C shows a poorer crystallinity while a better capacity of 302 mAh/g in the range of 1.8 - 4.0 V, and its capacity remained 278 mAh/g after 30 cycles. Meanwhile, A simple and low-cost solvothermal method has been developed to synthesize γ-LiV_2O_5 nanorods by using ethanol as a solvent as well as a reducing agent. Preliminary electrochemical tests indicated that the γ-LiV_2O_5 obtained at 160 °C has an initial specific capacity of 259 mAh/g in the range of 1.5–4.2 V. It indicates that the lithium vanadium oxide nanorods prepared by the above methods have potentiality to be used as cathode material in rechargeable lithium batteries.

REFERENCE

[1] Panero, S; Pasquali, M; Pistoia, G. *J. Electrochem. Soc.*, 1983, 130, 1225.
[2] Pistoia, G; Pasquali, M; Wang, G; Li, LJ. *J. Electrochem. Soc.*, 1990, 137, 2365.
[3] Manev, V; Momchilov, A; Nassalevska, A; Pistoia, G; Pasquali, M. *J. Power Sources.*, 1995, 54, 501.
[4] West, K; Zachau-Christiansen, B; Skaarup, S; Saidi, Y; Barker, J; Olsen, II; Pynenburg, R; Koksbang, R. *J. Electrochem. Soc.*, 1996, 143, 820.
[5] Kumagai, N; Yu, A. *J. Electrochem. Soc.*, 1997, 144, 830.

[6] Delmas, H; Cognac-Auradou, JM; Cocciantelli, M; Menetrier, JP. Doumerc, *Solid State Ionics.*, 1994, 69, 257.
[7] Murphy, W; Christian, PA; Disalvo, FJ; Waszczak, JV. *Solid State Ionics.*, 1979, 18, 2800.
[8] Cocciantelli, JM; Menetrier, M; Delmas, C; Doumerc, JP; Pouchard, M; Broussely, M; Labat, J. *Solid State Ionics.*, 1995, 78, 143.
[9] Rozier, P; Savariault, JM; Galy, J. *Solid State Ionics.*, 1997, 98, 133.
[10] Dai, J; Li, SFY; Gao, Z; Siow, KS. *Chem. Mater.*, 1999, 11, 3086.
[11] Chirayil, T; Zavalij, P; Whittingham, MS. *Solid State Ionics.*, 1996, 84, 163.
[12] Chirayil, T; Zavalij, P; Whittingham, MS. *J. Electrochem. Soc.*, 1996, 143, L193.
[13] Xu, HY; Wang, H; Song, ZQ; Wang, YW; Zhang, YC; Yan, H. *Chem. Lett.*, 2003, 32, 444.
[14] Xu, HY; Wang, H; Song, ZQ; Wang, YW; Yan, H; Yoshimura, M. *Electrochimica Acta.*, 2004, 49, 349.
[15] Wang, YW; Xu, HY; Wang, H; Zhang, YC; Song, ZQ; Yan, H; Wan, CR. *Solid State Ionics.*, 2004, 167, 419.
[16] Wadsley, AD. *Acta Cryst.*, 1957, 10, 261.
[17] Pistoia, S; Panero, M; Tocci, R; Moshtev, V; Manev, J. *Solid State Ionics.*, 1984, 13, 12.
[18] Pistoia, M; Pasquali, M; Tocci, V; Manev, R; Moshtev, J. *Power Sources.*, 1985, 15, 13.
[19] Yu, A; Kumagai, N; Liu, Z; Lee, JY. *J. Power Sources.*, 1998, 74, 117.
[20] Liu, Q; Zeng, CL; Yang, K. *Electrochimica Acta.*, 2002, 47, 3239.
[21] Kawakita, T; Kato, Y; Katayama, T; Miura, T; Kishi, J. *Power Sources.*, 1999, 81-82, 448.
[22] Oka, Y; Yao, T; Yamamoto, N. *Mater. Res. Bull.*, 1997, 32, 1201.
[23] Evans, HT; Black, S. *Inorg. Chem.*, 1966, 10, 1808.
[24] Zhang, X; Frech, R. *Electrochimica Acta.*, 1998, 43, 861.
[25] Anderson, DN; Willett, RD. *Acta Crystallogr. Sect. B.*, 1971, 27, 1476.
[26] Zhang, XL; Frech, R. *Electrochimeca Acta.*, 1997, 42, 475.
[27] Wagner, CD. *Handbook of X-ray Photoelectron Spectroscopy*, Minnesota: Perkin-Elmer Corporation, 1979.

INDEX

A

B

C

D

G

H

I

J

K

L

M

N

O

P

S

W

X

Y

Z